Elementary Principles of Flow for Hydraulic Engineers

By

Hunter Rouse

Copyright © 2011 Read Books Ltd.
This book is copyright and may not be
reproduced or copied in any way without
the express permission of the publisher in writing

British Library Cataloguing-in-Publication Data
A catalogue record for this book is available from
the British Library

ELEMENTARY PRINCIPLES OF FLOW

Velocity of a Fluid Particle as a Function of Time and Space. Any fluid may be imagined to consist of innumerable small but finite particles, each having a volume so slight as to be negligible when compared with the total volume of the fluid, yet sufficiently large to be considered homogeneous in constitution. Since these fluid particles must be in constant contact with each other, true impact between particles is physically impossible; it follows that any relative motion within the fluid will generally involve both rotation and deformation of each individual particle.

Each particle at any instant of time will have its own particular velocity, which will generally vary as it travels from point to point; moreover, the velocity of successive particles passing a fixed point is likewise a variable in the most general case. The Lagrangian method of attack studies the behavior of a given fluid particle during its motion through space; opposed to this is the method of Euler, which observes the flow characteristics in the immediate vicinity of a given point as the particles pass by. While perhaps not so descriptive of the fate of the individual particle, equations of motion obtained by the Eulerian method lend themselves more readily to practical use.

Unlike scalar quantities, such as length or time, velocity and acceleration may vary in direction as well as in magnitude; they are, therefore, true vectors, with components in each of three coordinate directions. Variation with time of any vector quantity may result from a change either in direction or in magnitude —or in both—of the vector itself. However, knowledge of vector analysis is not essential to the study of fluid motion, for the variation of a vector may be fully described by the changes in magnitude of its three components.

What is known in hydromechanics as a stream line is an imaginary curve connecting a series of particles in a moving fluid in such manner that at a given instant the velocity vector of every particle on that line is tangent to it (Fig. 1). In uniform flow the magnitude and direction of the velocity vector are constant over the entire length of any stream line at any instant; in other words, all stream lines remain parallel to one another, the distinction between uniformity and non-uniformity referring specifically to the geometry of the flow pattern. On the other hand, regardless of the relative form of the stream lines, if the flow is steady there may be no variation with time in either magnitude or direction of the velocity vector at any stationary point in the space through which the fluid moves.

It follows that a flow which is non-uniform and unsteady shows velocity variation both with distance and with time; that is, the velocity of fluid particles changes from point to point, and the velocity of the particles passing any given point changes from instant to instant. This is the most general case, and may be described mathematically by the expression

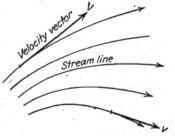

Fig. 1.—Stream lines.

$$v = f(t, x, y, z)$$

which states simply that the velocity vector is a function of time and of position with respect to the three coordinate axes. It is evident that in unsteady, non-uniform motion the entire flow pattern may be changing from instant to instant, in which case the stream lines must be regarded as instantaneous—that is, they represent the paths of particles for only a small increment of time. If the flow is either steady or uniform, the stream lines will represent the actual paths of the fluid particles; only in regions of uniform flow can these paths be parallel.

The same distinction that exists between a fluid particle and a point in the fluid medium may be used to define a stream filament as distinguished from a stream line: The reader must visualize a small filament or tube of fluid, bounded by stream lines and yet of inappreciable cross-sectional area, as shown schematically in Fig. 2. This stream filament might be considered, in either steady or uniform flow, as the passage through space of a fluid particle, and as such is the basis of the one-dimensional treatment of certain flow problems. Indeed, elementary hydraulics is based largely upon this conception, a single filament being assumed to have the cross-sectional area of the entire flow.

In certain types of fluid motion the stream filaments are arranged in a very orderly fashion, and may be made visible experimentally by the introduction of colored fluid at some point in the flow. More generally, however, there occurs a complex interlacing of the actual stream lines; the various particles not only follow completely different and intricate courses but suffer continuous distortion and subdivision, so that no particle exists as an individual for more than a short interval of time. In such cases it is often practicable to represent by stream lines or filaments the temporal average of conditions throughout the movement. Such representation does not ignore the actual complexity of the motion, but serves only as a convenient aid in visualizing the underlying pattern of the flow.

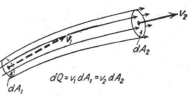

FIG. 2.—Stream filament.

The boundaries of fluid motion are the fixed or moving surfaces which define the borders of the fluid medium. These may either surround the flow, as is done by the walls of a pipe or turbine, or may be enclosed by the flow, as in the case of an airplane or the blade of a turbine runner. On the other hand, the free surface of a liquid in contact with a gas is not given its form by a solid boundary, but through the condition that the pressure intensity at every point on the free surface must be the same. In any case, the situation is fully described by what are known as boundary conditions. It should be clear that the borders of the flow are always stream lines, since by definition and physical fact, respectively, the flow cannot cross either a stream line or the boundary line of motion.

Relationship of the Velocity Fields for Steady and Unsteady Flow. In general hydromechanics it is customary to distinguish between absolute and relative motion. For the purpose of this text, however, it will be more feasible to disregard the absolute space of the theoretician, and simply place emphasis upon the relation of the flow picture either to the fluid itself, or to a fixed or moving boundary. In order to clarify this principle, consider the two cases of a boat moving through still water, and water flowing around a bridge pier. The coordinate system may be represented by the quadrille-ruled ground glass of a camera which is suspended some distance above the water. If the camera is "related" (fixed) to the boat or to the bridge pier, the flow picture will be a steady one, and the stream lines will also

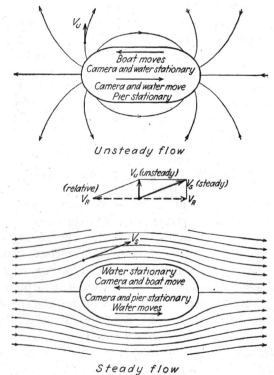

FIG. 3.—Patterns of relative motion between a fluid and a solid boundary.

be the apparent paths of the fluid particles (see Fig. 3). If, however, the camera is "related" to the water, the picture will change with time and will show the flow pattern caused by the pier or the boat at any instant. In the case of the bridge pier the water obviously moves through space, while in the other example it is the boat which moves; hence, in the first case the moving coordinate system gives a picture of unsteady motion, and in the second a picture of steady motion.

The relation between the two pictures is purely a vectorial one, as indicated in Fig. 3. Assume, for instance, that the camera is stationary above the water, thereby yielding the unsteady pattern caused by the boat moving to the left. If the camera is now moved to the left with the boat, the resulting steady pattern will be the same as if the boat and all fluid particles in the unsteady pattern were moved to the right at the relative velocity V_R. Thus the steady vector V_S at any point is the vector sum of V_R and the unsteady vector V_U. Similarly, the steady pattern of flow around the bridge pier, observed when the camera is stationary, may be changed to an unsteady one by moving the camera to the right—*i.e.*, by giving the pier and all fluid particles an apparent velocity V_R to the left. V_U is then the vector sum of V_R and V_S. Interesting photographic studies of such relative motion caused by a body moving through a fluid may be made by sprinkling aluminum flakes or bits of finely divided paper on the fluid surface and then making short time exposures: first while the camera moves at the exact speed of the body (steady motion), and then with the camera held motionless (unsteady motion). Such photographs may be seen in Fig. 4.[1]

A method of changing the picture of unsteady flow through a turbine or centrifugal pump, as would ordinarily be seen by an observer, to one of steady flow has been developed by Professor Thoma, of Munich. An apparatus known as a rotoscope, consisting of a small telescope and an objective prism, is mounted in line with the pump or turbine axis, the end of the casing having been replaced by a small section of plate glass. So long as the prism is at rest, the flow picture remains unsteady; but if the prism is made to rotate at the proper speed, through the telescope the blades will appear to be stationary. If dye is introduced at various points of the flow, a steady picture of the stream lines will result. Similar stroboscopic methods are in common use in studying the performance of high-speed machinery, involving the same principle of reducing unsteady motion to steady motion by relating the coordinate system to the moving parts.

It is customary to construct a stream-line diagram in such a fashion that the distance Δn between stream lines at every point

---

[1] For the purposes of simplicity, the stream lines in Figs. 3 and 5 have been drawn symmetrically at the front and rear of the body. That this is not an impossible case of flow is shown by the photographs, taken shortly after movement began. Further discussion of flow in the wake of a body must be reserved for a later chapter.

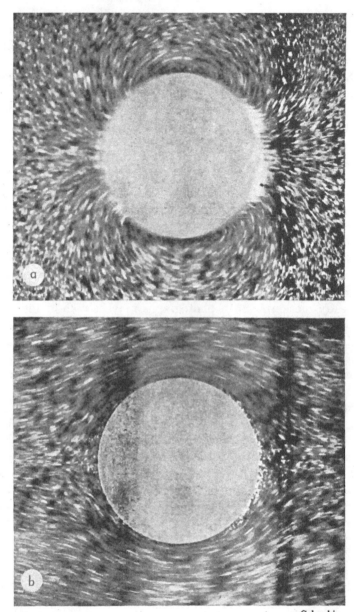

Columbia

Fig. 4.—Patterns of (a) unsteady and (b) steady flow around a cylinder as motion begins, the camera traveling with the fluid and with the cylinder, respectively. The stream lines are shown by the movement of highly illuminated aluminum particles during a short time exposure.

is inversely proportional to the length of the velocity vector at that point: $\Delta n = c/v$; this practice is possible on paper, of course, only in the case of two-dimensional or planar motion, such as that shown in Fig. 3. Because of such systematic selection of stream lines (in reality an infinite number of them exists), it is possible to simplify the construction of the pattern for steady flow from that of unsteady flow, and vice versa, with the following graphical method. For every system of flow there exists at each point of the pattern one and only one vector diagram representing the relation of the steady and unsteady velocity vectors to the velocity of translation; similarly, the magnitude of each vector represents a definite spacing of the stream lines. Since the relative velocity is constant, it may be represented by a series of equidistant lines parallel to the direction of motion,

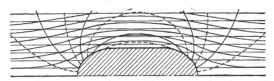

Fig. 5.—Graphical combination of velocity fields.

spaced according to the velocity and the proportionality constant. By connecting with smooth curves all points of intersection of these parallels with successive stream lines in either the steady or unsteady flow pattern, the corresponding pattern of flow lines will result, as shown in Fig. 5. This method of graphically combining (adding or subtracting) two vector fields is of especial value in the study of motion within the moving runner of a turbine, and will also be used in the following pages in the study of vector fields of force.

While the stream lines in steady flow represent the actual paths of fluid particles (with reference to the coordinate system), the paths traveled by particles in unsteady motion must, in general, be obtained by another graphical method. Since a given fluid particle in unsteady motion follows one stream line for only an infinitesimal increment of time, its velocity then being determined in direction and magnitude by the next stream line to cross its path, one must plot on a diagram of instantaneous stream lines the successive distances over which it will move in the direction of each stream line as the unsteady field of motion advances.

Thus, in Fig. 6, a particle at point 1 will move the distance a according to the direction and spacing of the stream line A at that point, the distance b according to conditions at point 2, and so on. The complete paths followed by the various particles in this particular system of motion will vary only with the original distance from the centerline of flow.

Stream lines and path lines must not be confused with so-called "streak" lines—the latter connecting all particles passing through a given point. Filaments of dye injected into a moving fluid therefore correspond to streak lines, and indicate path lines and stream lines as well only if the flow is steady.

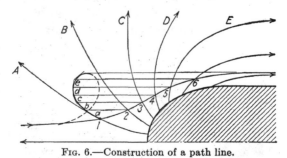

Fig. 6.—Construction of a path line.

Equations of Motion in Natural Coordinates. While the Cartesian coordinate system is by far the most general, it is at the same time very difficult to visualize states of motion described in this system, and just as difficult to adapt its equations of motion to various boundary conditions. Hence, it is customary to make extensive use of some specialized system, according to the case in question. In the study of flow through or around boundaries generally symmetrical about one axis, such as the surfaces of revolution in centrifugal pumps and turbines, the cylindrical coordinate system is particularly advantageous. The coordinates are again three in number, and consist of the radius r and the angle θ of the planar system of polar coordinates, and of the distance z measured along the axis of rotation. For very special cases the spherical (astronomical) system has its advantages, although its use is limited.

Because of its close relation to the stream-line picture, the natural coordinate system will be found particularly applicable to the methods of study in this book. This system is somewhat different from the others, in that it may be used to describe most conveniently only that region of flow immediately surrounding the center of coordinates. The coordinate center is thus located at the point in the flow chosen for study (point o in Fig. 7), the s axis being tangent to the stream line at that point, the n axis normal to the stream line and containing its center of curvature, and the m axis normal to the other two. It is apparent that at the coordinate center the velocity vector lies along the s axis, there being no velocity components at that point in either of the other coordinate directions. At any other point q, however (refer to Fig. 7), according to the coordinate axes centered at o, the velocity vector will have components in all three directions.

Writing the velocity at any point of an unsteady, non-uniform flow as a function of time and of position with respect to the three natural coordinate axes

$$v = f(t, s, n, m)$$

it will be seen that the velocity of a fluid particle must then vary not only according to the change of conditions in that locality with time, since the flow is unsteady, but also according to the change of conditions in each of the coordinate directions, since the flow is non-uniform.

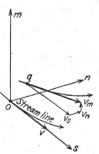

FIG. 7.—Characteristics of the natural coordinate system.

Since the velocity is a true vector, it will generally have components in each of the three coordinate directions (v_s, v_n, and v_m), each of which is also a function of time and space:

$$v_s = f_1(t, s, n, m)$$
$$v_n = f_2(t, s, n, m)$$
$$v_m = f_3(t, s, n, m)$$

Since acceleration is a vector quantity defined as the temporal rate of change of the velocity vector,

$$a = \frac{dv}{dt}$$

its components in the three coordinate directions may be written as the temporal rates of change of the corresponding components of velocity:

$$a_s = \frac{dv_s}{dt}; \quad a_n = \frac{dv_n}{dt}; \quad a_m = \frac{dv_m}{dt}$$

In moving an increment of distance in an increment of time, a particle will undergo acceleration for two distinct reasons: first, because of passage of time, without respect to its movement through space; second, because of its movement through space, without respect to passage of time. In other words, each component of total acceleration will consist of two parts: the first is written simply as the partial derivative of the velocity component with respect to time; the second is expressed as the partial derivative of the velocity component with respect to distance in the direction of motion, multiplied by the distance traveled per unit of time during the short time increment. Evidently the distance traveled per unit of time is equal either to the velocity component in the s direction or to the magnitude of the velocity vector itself, since the two differ by an exceedingly infinitesimal amount even after the particle has moved the short increment of distance. Hence, the three components of acceleration may be written in the following form:

$$a_s = \frac{dv_s}{dt} = \frac{\partial v_s}{\partial t} + \frac{\partial v_s}{\partial s}\frac{ds}{dt} = \frac{\partial v_s}{\partial t} + v\frac{\partial v_s}{\partial s} \tag{1s}$$

$$a_n = \frac{dv_n}{dt} = \frac{\partial v_n}{\partial t} + \frac{\partial v_n}{\partial s}\frac{ds}{dt} = \frac{\partial v_n}{\partial t} + v\frac{\partial v_n}{\partial s} \tag{1n}$$

$$a_m = \frac{dv_m}{dt} = \frac{\partial v_m}{\partial t} + \frac{\partial v_m}{\partial s}\frac{ds}{dt} = \frac{\partial v_m}{\partial t} + v\frac{\partial v_m}{\partial s} \tag{1m}$$

The partial derivatives express the local and the convective components of acceleration of the fluid particle, as distinguished from the total or substantial component of acceleration, and represent, respectively, variation with time, regardless of space, and with space, regardless of time. If the flow is steady, the local terms are zero; similarly, if the flow is uniform, the convective terms are zero. In a steady, uniform flow there is no acceleration whatsoever.

Owing to the nature of this coordinate system, these equations may be further simplified, for at the instant the fluid particle passes the origin o (Fig. 7) there can be no component of the velocity in either the n or the m direction, and the curvature of the stream filament will lie entirely in the sn plane over a short distance ds (i.e., the curvature of a line can be only two-dimensional at any point and hence over an infinitesimal distance on either side of that point). In Eq. (1s) the convective term may now be written in the following form, which involves merely a change in mathematical wording:

$$v\frac{\partial v}{\partial s} = \frac{\partial(v^2/2)}{\partial s}$$

The convective term in Eq. (1n) may be replaced by the familiar term of mechanics representing the centripetal acceleration of any mass moving in a curved path. This may be proved by reference to Fig. 8, similarity of triangles resulting in the proportion

$$\frac{\partial v_n}{\partial s} = \frac{v}{r}$$

whence the convective term in the second equation becomes simply v^2/r. Since the plane of curvature of the filament is normal to the m axis over the distance ds, there can be no acceleration in the direction m as the particle travels this short distance. Hence, the convective term in Eq. (1m) may be omitted entirely. These three equations may now be written as follows for the immediate vicinity of the coordinate center:

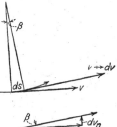

FIG. 8.—Centripetal acceleration.

$$\frac{dv_s}{dt} = \frac{\partial v_s}{\partial t} + \frac{\partial(v^2/2)}{\partial s} \tag{2s}$$

$$\frac{dv_n}{dt} = \frac{\partial v_n}{\partial t} + \frac{v^2}{r} \tag{2n}$$

$$\frac{dv_m}{dt} = \frac{\partial v_m}{\partial t} \tag{2m}$$

In order to study the individual action of any force property in producing such acceleration, it is necessary to eliminate, for the time being, the influence of all other force properties. This is accomplished in fundamental hydromechanics by arbitrarily setting the viscosity, surface tension, and compressibility of the fluid equal to zero, weight and pressure then being the forces under investigation. Thus, the forces exerted in an axial direction upon an elementary cylinder of fluid (see Fig. 9) will be the pressure at either end and the component of fluid weight acting parallel to the axis. While the pressure intensity at any point within such a fluid is the same in every direction, it will in general vary from point to point, its rate of variation in any direction being called the pressure gradient. The difference in pressure intensity between the two ends of the fluid cylinder is thus given by the pressure gradient in the axial direction times the distance

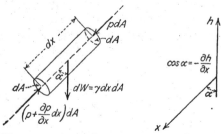

FIG. 9.—Elementary forces due to pressure gradient and weight.

between the two ends. The total force acting upon this fluid volume will then be:

$$dF_x = p\,dA - \left(p + \frac{\partial p}{\partial x}dx\right)dA + \gamma\,dx\,dA\,\cos\alpha$$

Introducing the rate of change of elevation h in the x direction ($\cos\alpha = -\partial h/\partial x$) this becomes:

$$dF_x = -\frac{\partial p}{\partial x}dx\,dA - \gamma\frac{\partial h}{\partial x}dx\,dA$$

This very important relationship may be expressed as follows: The force per unit volume, f, acting in any direction is equal to the rate of decrease of the sum $(p + \gamma h)$ in that direction:

$$\frac{dF_x}{dx\,dA} = f_x = -\frac{\partial}{\partial x}(p + \gamma h) \tag{3}$$

This force per unit volume divided by the density of the fluid will equal the force per unit mass, or, in accordance with the Newtonian equation, the rate of acceleration of the fluid in the given direction:

$$a_x = \frac{f_x}{\rho} \tag{4}$$

$$a_x = \frac{dv_x}{dt} = -\frac{1}{\rho}\frac{\partial}{\partial x}(p + \gamma h) \tag{5}$$

From Eq. (5) it will be seen that if any component of the substantial acceleration is zero, there can be no variation in the sum $(p + \gamma h)$ in that direction. In other words, the distribution of pressure intensity must be hydrostatic (*i.e.*, $p = $ constant $-\gamma h$) in any direction in which no acceleration takes place.

Equation (5) may now be combined with the expressions for acceleration already developed:

$$\frac{\partial v_s}{\partial t} + \frac{\partial (v^2/2)}{\partial s} = -\frac{1}{\rho}\frac{\partial}{\partial s}(p + \gamma h) \tag{6s}$$

$$\frac{\partial v_n}{\partial t} + \frac{v^2}{r} = -\frac{1}{\rho}\frac{\partial}{\partial n}(p + \gamma h) \tag{6n}$$

$$\frac{\partial v_m}{\partial t} = -\frac{1}{\rho}\frac{\partial}{\partial m}(p + \gamma h) \tag{6m}$$

These are special forms of the most basic relationships in hydromechanics, first published in 1755 by the founder of the science, the Swiss mathematician Leonhard Euler, and generally known as the Euler equations of acceleration.

12. Principles of Energy, Continuity, and Momentum. Equation (6s) may be written directly in the following pertinent form:

$$\rho\frac{\partial v_s}{\partial t} + \frac{\partial}{\partial s}\left(\frac{\rho v^2}{2} + p + \gamma h\right) = 0$$

The three terms within parentheses may be set equal to the quantity E_v. If the flow is assumed steady, the local acceleration will then be equal to zero, and integration of this differential expression over s will yield a general statement of conditions of steady flow along any stream line:

$$\int^s dE_v = \rho\frac{v^2}{2} + p + \gamma h = f(t) \tag{7}$$

Although the flow was expressly made a steady one, Eq. (7) states that the quantity E_v may still be a function of time. The reader will recall, however, that from definition the word "steady" specifies only that the velocity must remain constant with time at all points in the flow, and thereby places no restriction upon temporal variation of the pressure intensity. Equation (7) simply means that variation in the hydrostatic load on the system, as included in the sum $(p + \gamma h)$, will exactly equal the change in E_v with time and will extend uniformly over the entire length of the stream line; as such, it can have no effect whatever upon the velocity at any point.

If E_v is not a function of time, along any stream line

$$E_v = \rho \frac{v^2}{2} + p + \gamma h = \text{constant} \qquad (8)$$

Each term of this equation has the dimension of energy per unit volume, the equation embodying a complete statement of the energy principle, or the essential balance between kinetic energy and potential energy over every part of a stream line in steady flow. The equation of energy is more commonly known as the Bernoulli theorem, named for Daniel Bernoulli (like Euler a Swiss mathematician), who discussed the various forms of flow energy in a treatise on hydraulics (1738) nearly two decades before Euler laid the foundations of hydromechanics. The equation appears above in its most general form—one particularly well adapted to flow that is entirely confined by solid boundaries. Judging from previous remarks and from the interrelation of the three terms in this equation, it is evident that the pressure intensity will vary along the stream line with change in velocity, with change in elevation, and (when E_v is a function of time) with change in the hydrostatic load on the enclosed system. Under such conditions it is appropriate to distinguish that portion of the pressure intensity resulting from dynamic effects from that resulting from hydrostatic conditions. Designating these by the subscripts d and s, respectively, Eq. (8) may be written in the form

$$E_v = \rho \frac{v^2}{2} + p_d + p_s + \gamma h \qquad (9)$$

in which the sum $(p_s + \gamma h)$ is either a constant or some function of time; in either case,

$$\rho \frac{v^2}{2} + p_d = \text{constant} \qquad (10)$$

If Eq. (8) is divided by the fluid density ρ, there will result an alternate expression in which each term has the dimension of energy per unit mass of fluid:

$$E_m = \frac{v^2}{2} + \frac{p}{\rho} + gh \qquad (11)$$

This general form is particularly significant when dealing with the flow of gases. A form more familiar to engineers, since it is appropriate in cases of flow with a free surface, is derived from Eq. (8) by dividing all terms by the specific weight of the fluid:

$$E_w = \frac{v^2}{2g} + \frac{p}{\gamma} + h \tag{12}$$

Each term now has the dimension of energy per unit weight of fluid; since this is equivalent to length, the several terms are characterized as heads, and are called, respectively, the total head, the velocity head, the pressure head, and the geodetic head or elevation. Since the pressure head and elevation represent potential energy, as distinguished from the kinetic energy embodied in the velocity head, the sum $\left(\frac{p}{\gamma} + h\right)$ is properly known as the potential head. It follows from Eq. (8) that the sum of velocity and potential heads will not vary with distance along any stream line in steady flow. Evidently, no restriction is placed upon variation from one stream line to another.

At every point along a fluid surface exposed to the atmosphere, the pressure intensity must be that of the atmosphere itself. Hence, while the hydrostatic load in a closed system may be varied at will without changing the flow pattern, there is no possible way of changing the pressure intensity in flow with a free surface (except through atmospheric variation) without altering the entire pattern of motion. Under such circumstances differentiation between hydrostatic and dynamic influence upon the pressure intensity would be quite pointless, for the two can no longer be considered independent of each other.

Equation (12) finds a very valuable application in the graphical representation of the total head and its component parts. For a given stream filament in steady flow, the magnitude of its elevation at every point is plotted as a vertical distance above some assumed geodetic datum. The magnitude of the pressure head is then added vertically to the elevation of the stream line, the locus of these latter points being called the pressure line. From each point on the pressure line the corresponding velocity head is then laid off, resulting in a line of total head—the energy line—lying the distance E_w above the geodetic datum. Thus, in a single diagram is contained the entire story of energy transformation undergone by a fluid particle as it moves from point to point along its path, the method being quite as applicable to confined as to open flow.

If all neighboring filaments happen to possess the same total head, it is only reasonable to allow the same energy line to apply to the entire group. But since this in no way prevents pressure intensity and velocity from varying from one stream line to another, there will evidently be cases in which a different pressure line would have to be drawn for each individual filament; such is always the case if the stream filaments display appreciable curva-

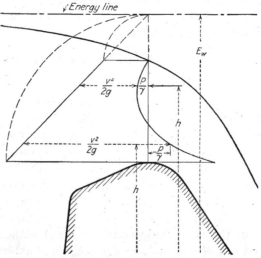

Fig. 10.—Non-hydrostatic pressure distribution in curvilinear motion.

ture. For illustration, in Fig. 10 is shown a longitudinal section through the crest of a spillway. Each filament passing the vertical line erected at the topmost point of the crest will be under a different pressure intensity, the pressure head at each elevation being plotted horizontally from the section line. Since the distance between the filament and the energy line is equal to the sum of pressure and velocity heads, it is obvious that the pressure line for every filament would lie a different distance below the energy line. As a matter of fact, only in parallel flow will a single pressure line suffice for all filaments at once: if the upper surface is exposed to the atmosphere, the pressure line will then coincide with the free surface; if the flow is confined, the pressure line may lie either above or below the centerline of flow, depending upon whether the mean pressure head is positive or negative (*i.e.*, greater or less than atmospheric).

In the most general case, however, one may not assume that all stream filaments possess the same energy of flow, for the total head often varies from one filament to the next. (It will be recalled that the Bernoulli theorem makes no mention of the relative energy of neighboring filaments.) Under such conditions, if one is still to use a common energy line for the entire flow, it is essential that the total head which this indicates be a true measure of the mean flow energy. In other words, since head represents energy per unit weight of fluid passing a given section, the total head for each increment of cross-sectional area must be weighted according to the rate of weight discharge through that elementary area.

Designating by $\Delta Q = v \Delta A$ the rate of discharge through the increment of cross-sectional area, the product of this quantity and the specific weight of the fluid will yield the desired weight passing the increment of area per unit time, and thus give the factor by which the total head of each filament must be multiplied before integrating over the entire cross section of flow. This integral, when divided by the mean rate of weight discharge, $\gamma V A$ (based on the mean velocity V), will equal the weighted mean total head:

$$(E_w)_m = \frac{1}{\gamma V A} \int^A \left(\frac{v^2}{2g} + \frac{p}{\gamma} + h \right) \gamma\, v\, dA \tag{13}$$

If the flow is parallel, on the other hand, the sum of pressure head and elevation—that is, the potential head—must be a constant over any cross section, so that only the velocity head will then vary from one filament to the next. Under such circumstances, a weighted mean velocity head must be added to the common potential head to determine the weighted mean total head. The procedure is similar to the foregoing one:

$$\left(\frac{v^2}{2g} \right)_m = \frac{1}{V A} \int^A \frac{v^2}{2g} v\, dA = K_e \frac{V^2}{2g} \tag{14}$$

The kinetic-energy correction factor K_e will then be:

$$K_e = \frac{1}{V^3 A} \int^A v^3\, dA \tag{15}$$

It may vary in magnitude from unity, for uniform total head, through an average practical value of about 1.1, to a general maximum of 2 when the velocity distribution (over a circular section) is paraboloidal. Needless to say, if the velocity distribution is modified by a change in flow section, the magnitude of this correction factor cannot remain constant.

In addition to the principle of energy, which is really the law of conservation of energy applied to the stream filament, the law of conservation of matter plays an essential role in hydromechanics. Since from definition there can be no passage of fluid through the walls of a stream filament (and since there may be no change in fluid density from one point to the next), unless the cross-sectional area changes with time, the rate of discharge must be the same at all cross sections of the filament at any instant:

$$v \, dA = \text{constant}$$

If one integrate this increment of discharge through the individual filament over the entire cross section of flow, the same conditions of continuity must hold through the flow at any one instant of time:

$$Q = \int^A v \, dA = \text{constant} \tag{16}$$

This equation applies to both steady and unsteady flow so long as the outermost stream lines do not change in form.

The Euler equations of acceleration, from which the energy equation was derived, embody the application of Newton's momentum principle to a particle at a given point in a moving fluid. With little modification, this principle may be applied conveniently to an entire region of flow, by integrating over the total volume of that region the elementary forces producing mass acceleration. The basic vector relationship between the force per unit volume and the rate of change of momentum

$$f = \frac{d(\rho v)}{dt} = \rho \frac{dv}{dt} = \rho a$$

may best be integrated over a given volume V by reducing the vector terms to their components in the three Cartesian coordinate directions; thus, the integral expression for any direction x will be:

$$F_x = \Sigma\,(f_x) = \rho \int^V a_x\,dV$$

The term on the left includes the x component of all forces acting upon every particle in the volume at a given instant. But since every force upon a particle within the volume requires the existence of an equal and opposite force upon the neighboring particles, all such internal forces will counterbalance each other, so that one need consider only those forces exerted externally. In its most general application, this quantity must include every type of force action; for the present, only pressure and weight are to be considered.

In the case of steady motion, the term on the right of the equation may be made more explicit by considering the fluid volume to be composed of innumerable fluid filaments of permanent form, the surface of this volume then consisting of the walls of the outermost filaments and the sum of all the cross-sectional areas dA at either end of every one. The component of acceleration, a_x, may now be expressed in terms of a differential length, ds, along any filament:

$$a_x = \frac{dv_x}{dt} = \frac{\partial v_x}{\partial s}\frac{ds}{dt} = v\,\frac{\partial v_x}{\partial s}$$

Since the differential volume dV is equal to the product $ds\,dA$, and since the term $v\,dA$ is equal to an increment of discharge dQ (from the equation of continuity), the term on the right of the original equation finally becomes:

$$\rho \int^s \int^A v\,\frac{\partial v_x}{\partial s}\,ds\,dA = \rho \int^s \frac{\partial v_x}{\partial s}\,ds \int^A v\,dA = \rho \int^Q (v_{x_2} - v_{x_1})\,dQ$$

in which v_{x_1} and v_{x_2} denote the components of the velocity at entrance to and at exit from the given space.[1]

The equation of momentum applied to an appreciable section of the flow will then have, for any coordinate direction, this general form:

---

[1] This development, as well as many another feature of elementary hydromechanics, is clearly and simply presented in MISES, R. VON, "Technische Hydromechanik," B. G. Teubner, Leipzig, 1914.

$$F_x = \Sigma (f_x) = \rho \int^Q v_{x_2} \, dQ - \rho \int^Q v_{x_1} \, dQ \qquad (17)$$

When reduced to the simple form used in hydraulics, the two integral terms are replaced by the product of the rate of discharge and the average velocity components at entrance and exit. In this form the equation is commonly applied to cases of jets deflected by curved surfaces, to pipe bends, nozzles, and similar elementary hydraulic devices. Just as in the case of the total head, however, if the velocity is not uniformly distributed over the section, it is not correct simply to use the mean value V of the velocity in the momentum equation. From Eq. (17) it is evident that to the product $\rho V^2 A$ at each section there must be applied a correction factor of the form:

$$K_m = \frac{1}{V^2 A} \int^A v^2 \, dA \qquad (18)$$

This factor must not be confused with K_e, used in the energy equation, for the velocity appears to the second power in the former and to the third power in the latter.

In the general case of curvilinear flow, in which average values of neither velocity nor pressure intensity may be used, Eq. (17) must be followed strictly, actual curves of velocity and pressure distribution forming the basis for integration, and the actual volume of the fluid being used to determine the component of the fluid weight in the given direction. Yet such methods will require experimental measurement of velocity and pressure distribution at one section or the other, for the general principles of momentum, energy, and continuity have as yet provided no means of determining these characteristics by rational analysis.

Theory and Use of the Flow Net. If the total energy is constant not only along any stream filament but also from one filament to another, Eq. (6n) may be simplified, owing to the fact that the sum of kinetic and potential energy cannot vary in the n direction. Since the flow must necessarily be steady, the term $\partial v_n/\partial t$ can be dropped at once. Then subtracting the quantity $\dfrac{\partial(v^2/2)}{\partial n}$ from both sides of the equation,

$$\frac{v^2}{r} - \frac{\partial(v^2/2)}{\partial n} = -\frac{1}{\rho}\frac{\partial}{\partial n}\left(\frac{\rho v^2}{2} + p + \gamma h\right)$$

and since the flow is one of uniform energy distribution, the right side of this equation must equal zero; hence,

$$\frac{v^2}{r} = v \frac{\partial v}{\partial n} \quad \text{or} \quad \frac{\partial v}{\partial n} = \frac{v}{r} \tag{19}$$

This equation may be integrated to give

$$\ln v = \int \frac{dn}{r} + C \quad \text{or} \quad v = C' \, e^{\int \frac{dn}{r}} \tag{20}$$

In combination with the principle of continuity and the Bernoulli theorem, Eq. (19) affords a means of determining the pressure and velocity distribution in steady two-dimensional flow in which the total head is reasonably uniform at all points. While this method is necessarily graphical (with the exception of certain cases to be discussed later), it is nevertheless based upon physically exact principles, and is limited only by the degree of graphical accuracy which may be attained. Consider, for instance, the fixed boundaries shown in Fig. 11a. A flow net is first sketched in by eye, consisting of an arbitrary number of stream lines, and of such a number of orthogonal lines as to divide the flow area into as nearly perfect squares as possible, all stream lines and orthogonal lines meeting at right angles. It is evident that a network of diagonals through all points of intersection should also yield squares, and thus permit a convenient test for angular error. Obviously the squares in the vicinity of the bend will be distorted, but they should become more nearly perfect as the number of lines is increased. Mathematically speaking, they can become true squares only as they become infinitesimal in size. However, by adjusting the position of stream lines and orthogonal lines, with plentiful use of pencil and eraser, a very systematic progression of the stream lines will finally result.

There is only one arrangement of the chosen number of stream lines that will fulfill the mathematical statement of the equations of Euler; consequently, their positions may now be checked for correctness by means of Eq. (19). For any orthogonal line a plot is made of the relationship between distance along the orthogonal in the direction n and the radii of curvature of the stream filaments at the points of intersection with the orthogonal. These values must be determined graphically. From the principle of continuity as applied to the flow between each pair of stream lines (*i.e.*, the velocity must vary inversely with the spacing), the velocity vector at each point of intersection is next determined and plotted against n on the same diagram. If the stream lines are correctly spaced, it is evident from Eq. (19) that the slope of the velocity curve at any point must equal the ratio between the velocity and the radius of curvature, which can be checked

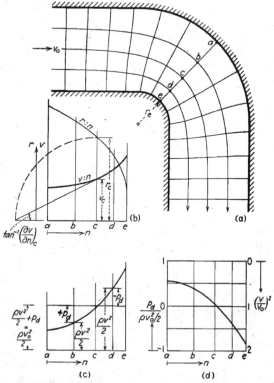

Fig. 11.—Application of the flow net.

graphically on the plot (see Fig. 11b). If these values do not agree as they should, the stream lines must be redrawn.

Usually a single check at several typical points will serve to make the diagram as accurate as is required. Once the correct velocity curve is obtained, the pressure distribution may be found through use of the energy equation (see Fig. 11c). Since the flow is enclosed, the change in pressure intensity with change in elevation (or with variation in hydrostatic load upon the system) can have no effect whatever upon the velocity distribution or upon the distribution of dynamic pressure. Indeed, if the pressure distribution were computed in terms of head, the variation in p_d/γ with n would be exactly equal to the variation in reading of manometric columns connected to piezometers at the several points in question.

Little thought will be required to convince the reader that the same flow net may apply equally well to rates of discharge other than that selected for investigation, for the pattern of stream lines depends not upon the velocity represented by the chosen spacing, but purely upon the geometrical form of the boundaries. Since a change in velocity will change the scale but not the form of the distribution curves in Fig. 11c, it would be most expedient to devise a single diagram containing the ratio of velocity, density, and dynamic pressure intensity for the given section, regardless of the rate of discharge (Fig. 11d). If this ratio is put in the form

$\dfrac{p_d}{\rho\, v_0^2/2}$, it will not only be dimensionless but will be identical

with the flow parameter derived by means of the Π-theorem in the preceding chapter. In other words, at any given point in the flow pattern, this parameter must remain constant regardless of change in rate of discharge or hydrostatic load—a fact of fundamental importance.

If the flow has one or more free surfaces, the problem becomes considerably more difficult, although no less subject to solution. If the profile of the free surface is known from experimental measurement, the procedure is the same as though this curve were one of the given boundaries; but if such experimental information is not available, since the stream lines at the border between liquid and atmosphere are not governed in position by a fixed boundary, they too must be sketched in by eye. The identical construction of the flow net then follows. However, in checking the position of the stream lines, not only must Eq. (19) be satisfied, but the pressure distribution curve must pass through zero (atmospheric) at every point lying on a free surface. If this does not check, not only the profile of the free surface, but the form of all other stream lines as well, must be modified.

As has already been mentioned, use of the energy equation written in terms of head is best suited to such conditions of flow. The scale used in plotting velocity and pressure head is then determined directly by the linear scale of the flow profile, and consequently the several distribution curves have immediate quantitative significance. For illustration, in Fig. 12 is shown the flow net for discharge under a sluice gate, the solution yielding curves of pressure head over the gate and along the lower boundary of the flow.

The gradual rise in the upper surface approaching the sluice gate illustrates a characteristic of flow in the neighborhood of a so-called point of stagnation. Wherever a stream line changes abruptly in direction, the spacing of the stream lines at that point is either infinity or zero, depending upon whether the change

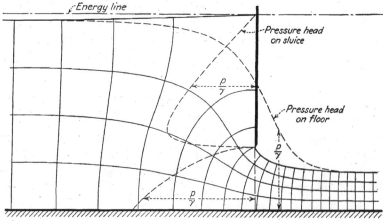

FIG. 12.—Pressure distribution for sluice-gate discharge as determined from the flow net.

in direction is toward or away from the other stream lines. Since, from continuity, the product of the spacing and the velocity must remain constant for a given rate of discharge, a zero spacing denotes an infinitely high velocity, and vice versa. Although a velocity equal to infinity (corresponding, in enclosed flow, to a dynamic pressure intensity of negative infinity) is physically impossible, a velocity of zero is obviously quite conceivable. If there are no free surfaces involved, this simply means that the intensity of dynamic pressure must increase to the magnitude of the constant in Eq. (10), and then be known as the stagnation pressure intensity. Indeed, it is upon this condition of stagnation that operation of the Pitot tube depends, for the nose of the tube pointing upstream merely produces a small region of stagnation at which the pressure intensity is measured; the ratio of stagnation pressure to the density and the square of the velocity of the surrounding flow must remain constant, regardless of change in magnitude of the velocity itself. However, if the stagnation point lies at a free surface, as in the case of the sluice gate, the pressure intensity must evidently remain atmospheric; since the total head along this upper stream filament must be the same at all points, it is clear that the decrease in velocity head must be accompanied by an increase in elevation of the filament, its highest position coinciding with the energy line itself at the point of stagnation.

Karlsruhe

FIG. 13.—Flow profile at a ventilated overfall.

Significance of the Force Potential. According to the Euler equations, two distinct components of the accelerative force per unit volume act upon fluid particles in two-dimensional flow: one (f_s) in the direction of motion, and the other (f_n) in the direction of the center of curvature of the stream filament; these force components are normal to the orthogonal lines and the stream lines, respectively, of the flow net, and are equal to the negative gradients of the sum ($p + \gamma h$) in the corresponding directions, in accordance with Eq. (3). On the other hand, the resultant force f per unit volume at any point might also be considered to have components in two other directions: one (f_p) normal to a line of constant pressure intensity, and the other (f_w) normal to a line of constant elevation. Figures 14 and 15, based upon flow conditions at the free overfall,[1] clearly indicate

[1] ROUSE, H., "Verteilung der hydraulischen Energie bei einem lotrechten Absturz," Oldenbourg, Munich, 1933.

this essential distinction, the two force parallelograms for the same point in the flow yielding the same vector f.

A force potential may be defined in hydromechanics as a quantity whose derivative in any direction equals the component of force per unit volume acting in that direction. From this

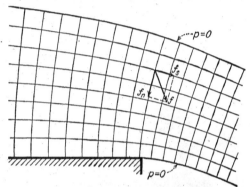

Fig. 14.—Velocity field at overfall crest.

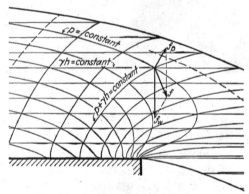

Fig. 15.—Force field at overfall crest.

definition it is evident that the force potential for flow under the action of weight and pressure must be the quantity $-(p + \gamma h)$, as will be seen from reference to Eq. (3). But since this expression may also be written in the form

$$f_x = \frac{\partial(-p)}{\partial x} + \frac{\partial(-\gamma h)}{\partial x}$$

it is clear that the quantities $-p$ and $-\gamma h$ also have the nature of force potentials; these may be called the pressure potential and the weight potential, respectively, since

$$(f_p)_x = -\frac{\partial p}{\partial x} \quad \text{and} \quad (f_w)_x = -\frac{\partial(\gamma h)}{\partial x}$$

The lines of constant pressure intensity shown in Fig. 15 thus represent lines of constant pressure potential, while the lines of constant elevation are lines of constant weight potential; in either case successive lines represent constant increments of potential. Hence, while the lines of weight potential are necessarily equidistant parallels, variation in pressure intensity from point to point in the flow results in a gradual change in form and spacing of the pressure-potential lines. Because of the constant increment in potential from line to line, at a given point the force component in any direction for either pressure or weight is inversely proportional to the spacing of the lines in that direction; evidently, it attains its maximum value when normal to the potential line passing through the given point.

Since either system of lines represents a vector field of force, the two systems may be combined vectorially to yield a resultant system corresponding to the vector field of the resultant force f. Moreover, inasmuch as a line of total force potential must represent a constant value of the sum $(p + \gamma h)$, such vectorial combination may be accomplished graphically simply by connecting with smooth curves (the heavy lines in Fig. 15) successive points of intersection of the two component systems. Since the total force per unit volume at any point must be normal to the resultant potential line passing through this point, it is evident that the convective acceleration produced by this field of force must also be normal to the potential lines at all points of the flow. In other words, while the flow net shows at a glance the direction and relative magnitude of the velocity vector, the pattern of the potential field indicates the direction and relative magnitude of the vector of force or acceleration.

It should be evident from the foregoing discussion that the actual magnitude of either the pressure or the weight potential is of no consequence, for the corresponding force component is determined entirely by the rate of change of the potential in the given direction. The components f_p and f_w may vary greatly with respect to one another in different regions of a given flow, although if the flow is uniform their vector sum must always be zero. They will approach this condition as a limit, for instance, a very great distance upstream from the crest of the free overfall, where the lines of constant pressure and of constant elevation practically coincide; f_w will then be directed downward and f_p upward, the resultant lines of constant force potential then being vertical and extremely far apart. Within the falling nappe, on the other hand, as the pressure intensity approaches the atmospheric, the pressure gradient will approach zero, so that the lines of constant force potential will become more nearly coincident with lines of constant elevation.

So long as a given state of steady unconfined flow is the immediate result of gravitational action, at any point in the flow

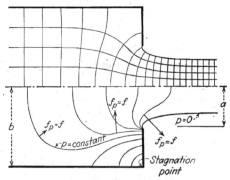

Fig. 16.—Velocity and force patterns for a two-dimensional orifice.

the ratio of f_p to f_w (i.e., the direction of f) is fixed once and for all by the existing boundary conditions. Variation in the fluid properties of density and specific weight will most certainly change the magnitudes of these components and of the velocity of flow as well, but the flow net and the lines of force potential cannot possibly change in form. Similar circumstances are encountered in the case of completely confined flow, with the exception that either velocity or the fluid properties may be varied independently, and thereby produce a change in magnitude —but not direction—of the acceleration at any point; since weight can produce no acceleration in confined flow, $f_w = 0$ and $f = f_{p_d}$.

Distinctly different is the case of flow that is confined over only a part of its course, a condition illustrated by efflux of a liquid from a closed conduit into the atmosphere. In Fig. 16, for instance, is shown the flow net for discharge from a two-dimensional orifice (i.e., from a very wide rectangular opening). So long as the velocity of efflux is very high, the magnitude of f_w is quite negligible in comparison with the high values of f_p in the neighborhood of the orifice, so that only as f_p approaches zero some distance from the orifice does the action of weight in deflecting the jet become appreciable. However, f_p may be varied at will simply by changing the pressure intensity within the conduit (thus producing a change in the total energy of flow), gradual lowering of which will make f_p and f_w more and more nearly of the same order. Under such conditions the deflection of the jet will become more and more appreciable in the neighborhood of the opening, the jet thereby losing its original symmetry.

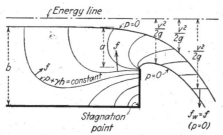

Fig. 17.—Force pattern for a sharp-crested weir.

The evident limit of such variation is reached when the flow occurs entirely as a result of gravitational attraction, under which condition the flow energy is a minimum for the given boundaries. Such a case is illustrated by the suppressed weir of Fig. 17, the proportions of which have been made to correspond to the lower half of the orifice of Fig. 16. Significant is the fact that the lines of constant force potential are identical some distance upstream in both cases, despite the fact that weight is of no moment whatever in the one; only in the unconfined portion of the flow do these lines differ appreciably in form, and even then one can readily visualize a systematic progression from one to the other with variation in the relative magnitudes of the two force components. The reader will realize that horizontal acceleration within the nappe becomes negligible only as the potential lines approach the horizontal; only then will the nappe assume the parabolic trajectory of free fall.

It has been shown that the relative magnitudes of f_p and f_w in cases of efflux may be varied over a great range through variation in the energy of flow. Though the fact is not generally appreciated, similar circumstances may be realized in weir flow as well. In the former instance, energy variation was accomplished by a change in pressure intensity within the conduit; with flow upstream from the weir freely exposed to the atmosphere, however, arbitrary change in pressure intensity is impossible. Yet the energy of flow may still be increased through increase in velocity head—accomplished, for instance, through discharge from a sluice gate or down a steep incline. Under such circumstances, it is evident from the momentum principle that the distribution of pressure intensity, and hence of pressure potential, could be altered almost at will with respect to a constant pattern of weight potential. Obviously, this would entail considerable variation in the form of the free surface, along which the pressure intensity must always remain atmospheric.

GENERALIZED EQUATIONS

Translation, Rotation, and Deformation of a Fluid Element.
For a more detailed study of the behavior of the individual
particle than is possible in natural coordinates, one must turn
to a more general coordinate system, the Cartesian. While the
method of Lagrange, in which the particle is followed along its
course, yields equations of definite significance, continued adher-
ence to the procedure originated by Euler of observing conditions
at fixed points in space will prove of greater value in the present
discussion; the behavior of particles will then be studied as they
pass the points in question.[1]

During the very small increment of time dt the total variation
of any velocity component v_x of a fluid particle must be equal
to its rate of change with time $\partial v_x/\partial t$ multiplied by the time
increment, plus its rate of change with movement in each of the
coordinate directions, $\partial v_x/\partial x$, $\partial v_x/\partial y$, and $\partial v_x/\partial z$, multiplied by
the increment of distance traveled in the corresponding direction
in the same increment of time:

$$dv_x = \frac{\partial v_x}{\partial t} dt + \frac{\partial v_x}{\partial x} dx + \frac{\partial v_x}{\partial y} dy + \frac{\partial v_x}{\partial z} dz$$

As before, the partial derivatives represent variation with time,
regardless of space, and with each direction in space, regardless
of time and of the other two coordinate directions.

Since the distance a particle travels in any direction during an
increment of time is equal to its component of velocity in this
direction multiplied by the time increment, the following relationships exist:

$$dx = v_x \, dt; \qquad dy = v_y \, dt; \qquad dz = v_z \, dt$$

---

[1] LAMB, H., "Hydrodynamics," 6th ed., pp. 2–6, 31–35, Cambridge
University Press, 1932.

From these three equations it is evident that the increments of distance traveled in the three directions in the same increment of time must be proportional to the velocity components in the respective directions; this proportionality is, in effect, the differential equation of the instantaneous stream line to which the velocity vector of the particle is tangent:

$$\frac{dx}{v_x} = \frac{dy}{v_y} = \frac{dz}{v_z} = \frac{ds}{v} \tag{21}$$

If the foregoing equivalents of dx, dy, and dz are substituted in the original expression, there will result:

$$dv_x = \frac{\partial v_x}{\partial t} dt + v_x \frac{\partial v_x}{\partial x} dt + v_y \frac{\partial v_x}{\partial y} dt + v_z \frac{\partial v_x}{\partial z} dt$$

Similar operations, of course, may be performed for each of the remaining coordinate directions. It is then obvious from inspection that division of each equation by the time increment will yield the total or substantial acceleration of the particle in each coordinate direction. From Eq. (5) these components of acceleration may then be equated to the component of force per unit mass acting upon the particle in the corresponding direction; thus are derived the three fundamental equations of Euler:

$$\frac{\partial v_x}{\partial t} + v_x \frac{\partial v_x}{\partial x} + v_y \frac{\partial v_x}{\partial y} + v_z \frac{\partial v_x}{\partial z} = -\frac{1}{\rho} \frac{\partial}{\partial x}(p + \gamma h) \tag{22x}$$

$$\frac{\partial v_y}{\partial t} + v_x \frac{\partial v_y}{\partial x} + v_y \frac{\partial v_y}{\partial y} + v_z \frac{\partial v_y}{\partial z} = -\frac{1}{\rho} \frac{\partial}{\partial y}(p + \gamma h) \tag{22y}$$

$$\frac{\partial v_z}{\partial t} + v_x \frac{\partial v_z}{\partial x} + v_y \frac{\partial v_z}{\partial y} + v_z \frac{\partial v_z}{\partial z} = -\frac{1}{\rho} \frac{\partial}{\partial z}(p + \gamma h) \tag{22z}$$

Consider now a fluid element of cubical form (shown in Fig. 18) having small but finite sides δx, δy, and δz, each parallel to the respective axis. At the point (x, y, z)—the corner of the cube nearest the origin—the velocity vector has the three components v_x, v_y, and v_z. Since the given state of flow varies with distance in each of the coordinate directions, at any instant the velocity components at every other corner of the cube will differ from these three by amounts depending upon the length of the sides and upon the gradient of each velocity component in each of the three coordinate directions.

In order to avoid the confusion of considering variations in all three directions at once, it will suffice to follow the changes in any one face of the cube, later extending the relationships thereby developed to the other faces. Taking, for instance, the face nearest the plane of the x and y axes, the velocity components in the directions x and y at the four corners of this face will be as indicated in Fig. 19. Neglecting for the moment the local varia-

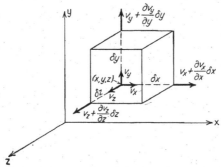

FIG. 18.—Variation in velocity along the boundaries of a fluid element.

tion (the change with time at a fixed point in space), it will be seen from the illustration that during a small time increment dt the actual motions of the several corners of the face must be different. Hence, not only will the face be moved bodily through space, but it must at the same time suffer a change in its original form.

In order to bring clarity into such a complex picture, let the motion be reduced to the four essential types of movement which the face may undergo: Superposed upon the translation of the square in the x and y directions, there will be in the most general case a change in the length of each pair of parallel sides (linear deformation), a change in each of the four corner angles (angular deformation), and a turning movement in one direction or the other (rotation). Each of these essential types of displacement is shown schematically in Fig. 20.

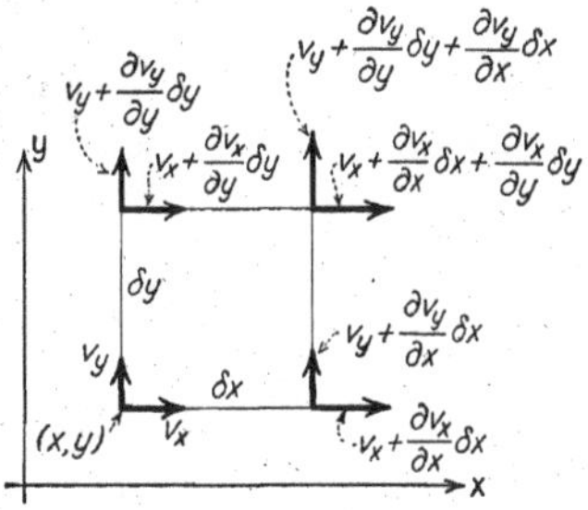

FIG. 19.—Velocity components in the xy plane.

During the time increment dt the magnitude of the translation in the two directions will be represented by the quantities

$$v_x\, dt \quad \text{and} \quad v_y\, dt$$

The magnitude of the linear deformation will be given by the difference between the distances moved by each pair of opposite sides:

$$\frac{\partial v_x}{\partial x}\,\delta x\, dt \quad \text{and} \quad \frac{\partial v_y}{\partial y}\,\delta y\, dt$$

Angular deformation, considering only the change in the right angle at the point (x, y, z), will depend upon the difference between the angular movements $d\alpha$ and $d\beta$ of the two sides δx and δy. Since over a very short time these angular increments

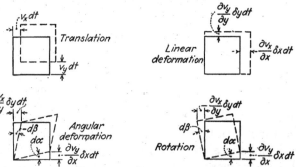

FIG. 20.—Schematic representation of translation, deformation, and rotation in the xy plane.

will be small, they may be considered numerically equal to their respective tangents; selecting the counterclockwise direction as positive:

$$d\alpha = \frac{\frac{\partial v_y}{\partial x}\,\delta x\, dt}{\delta x}; \quad d\beta = -\frac{\frac{\partial v_x}{\partial y}\,\delta y\, dt}{\delta y}$$

$$d\alpha - d\beta = \left(\frac{\partial v_y}{\partial x} + \frac{\partial v_x}{\partial y}\right) dt$$

Rotation, on the other hand, will occur if the angular increments are unequal or of like sign; the total angle through which the face is rotated will be equal to their average value:

$$\frac{d\alpha + d\beta}{2} = \frac{1}{2}\left(\frac{\partial v_y}{\partial x} - \frac{\partial v_x}{\partial y}\right) dt$$

It will be apparent that translation and rotation are allied types of displacement, since each denotes a bodily movement of the face without changing its original form. Similarly, linear and angular deformation bear a definite relationship to each other, since the former generally involves a change in the angles formed by the diagonals, whereas the latter involves a change in the diagonal lengths. Thus, linear deformation would become angular deformation according to axes turned through 45°, and vice versa.

Owing to the finite dimensions of the original cube, the values just given for each type of displacement are not exact, variables of order higher than the first having been omitted. But if the sides of the cube are now assumed to become infinitesimal, all corners then approaching the point (x, y, z) as a limit, each type of movement may be expressed exactly as a rate of change with time. At the given point the rate of translation in the three coordinate directions will be simply the three velocity components

$$v_x, \qquad v_y, \qquad v_z \quad \text{and} \quad v = \sqrt{v_x{}^2 + v_y{}^2 + v_z{}^2}$$

The rate of linear deformation in each of the coordinate directions will be

$$\frac{\partial v_x}{\partial x}\, dx, \qquad \frac{\partial v_y}{\partial y}\, dy, \qquad \frac{\partial v_z}{\partial z}\, dz$$

each denoting the velocity at which the respective opposite faces are drawing apart. Since the density (and hence the volume) of a fluid particle must remain constant during such deformation, it is obvious that an elongation of the particle in two directions must be compensated by a contraction in the third, and vice versa. Thus, the rates of linear deformation in the three coordinate directions (*i.e.*, the changes per unit time of the distances between opposite faces) multiplied by the areas of the respective faces must have a sum of zero:

$$\frac{\partial v_x}{\partial x}\, dx(dy\ dz) + \frac{\partial v_y}{\partial y}\, dy(dz\ dx) + \frac{\partial v_z}{\partial z}\, dz(dx\ dy) = 0$$

Dividing each term by the elementary volume yields the most general form of the equation of continuity, which states that the "divergence" of the velocity vector must equal zero in a flow in which the density is not a variable quantity; in other words, the velocity cannot increase or decrease in all three directions at once:

$$\text{div } v = \frac{\partial v_x}{\partial x} + \frac{\partial v_y}{\partial y} + \frac{\partial v_z}{\partial z} = 0 \qquad (23)$$

The rate of angular deformation at a given point (a quantity which varies with direction but which is not a true vector) is designated in each of the coordinate directions by the symbols ξ (xi), η (eta), and ζ (zeta). These symbols then represent the rate of angular deformation in planes normal to each of the three axes; for purposes which will be clear directly, they are arbitrarily made equal to one-half the actual rate of angular deformation:

$$\xi = \frac{1}{2}\left(\frac{\partial v_z}{\partial y} + \frac{\partial v_y}{\partial z}\right); \eta = \frac{1}{2}\left(\frac{\partial v_x}{\partial z} + \frac{\partial v_z}{\partial x}\right); \zeta = \frac{1}{2}\left(\frac{\partial v_y}{\partial x} + \frac{\partial v_x}{\partial y}\right) \quad (24)$$

On the other hand, the rate of rotation—angular velocity—is a true vector quantity and is given the customary symbol ω (omega); its components will then denote the rotation per unit of time in planes normal to the three axes, the sense of rotation determining the direction of the vector:

$$\omega_x = \frac{1}{2}\left(\frac{\partial v_z}{\partial y} - \frac{\partial v_y}{\partial z}\right); \omega_y = \frac{1}{2}\left(\frac{\partial v_x}{\partial z} - \frac{\partial v_z}{\partial x}\right); \omega_z = \frac{1}{2}\left(\frac{\partial v_y}{\partial x} - \frac{\partial v_x}{\partial y}\right) \quad (25)$$

The vector 2ω is known in hydromechanics as the curl of the velocity vector:

$$\omega = \sqrt{\omega_x{}^2 + \omega_y{}^2 + \omega_z{}^2} = \tfrac{1}{2} \text{ curl } v \qquad (26)$$

Since the convective acceleration at any point in space must depend not only upon the magnitude and direction of the velocity vector at the given point but also upon the divergence of the velocity vector and upon angular deformation and rotation, it should be possible to incorporate these quantities in the three equations of Euler. Thus the convective terms of Eq. $(22x)$ may be rewritten as follows, through simultaneous addition and subtraction of the proper quantities:

$$v_x\frac{\partial v_x}{\partial x} + v_y\frac{\partial v_x}{\partial y} + v_z\frac{\partial v_x}{\partial z} + \frac{1}{2}v_y\frac{\partial v_y}{\partial x} - \frac{1}{2}v_y\frac{\partial v_y}{\partial x} + \frac{1}{2}v_z\frac{\partial v_z}{\partial x} - \frac{1}{2}v_z\frac{\partial v_z}{\partial x}$$

$$= v_x\frac{\partial v_x}{\partial x} + \frac{1}{2}v_y\left(\frac{\partial v_y}{\partial x} + \frac{\partial v_x}{\partial y}\right) - \frac{1}{2}v_y\left(\frac{\partial v_y}{\partial x} - \frac{\partial v_x}{\partial y}\right)$$

$$+ \frac{1}{2}v_z\left(\frac{\partial v_x}{\partial z} + \frac{\partial v_z}{\partial x}\right) + \frac{1}{2}v_z\left(\frac{\partial v_x}{\partial z} - \frac{\partial v_z}{\partial x}\right)$$

On performing a similar operation upon each of the three components of convective acceleration and substituting the corresponding symbols from Eqs. (24) and (25), there will result:

$$\frac{\partial v_x}{\partial t} + v_x \frac{\partial v_x}{\partial x} + v_y \zeta + v_z \eta - v_y \omega_z + v_z \omega_y = -\frac{1}{\rho}\frac{\partial}{\partial x}(p + \gamma h) \quad (27x)$$

$$\frac{\partial v_y}{\partial t} + v_y \frac{\partial v_y}{\partial y} + v_z \xi + v_x \zeta - v_z \omega_x + v_x \omega_z = -\frac{1}{\rho}\frac{\partial}{\partial y}(p + \gamma h) \quad (27y)$$

$$\frac{\partial v_z}{\partial t} + v_z \frac{\partial v_z}{\partial z} + v_x \eta + v_y \xi - v_x \omega_y + v_y \omega_x = -\frac{1}{\rho}\frac{\partial}{\partial z}(p + \gamma h) \quad (27z)$$

Owing to the general character of the Cartesian system, it is seldom necessary to develop more than one of the three equations for any condition of flow, for the remaining two may always be found from the first through the method of cyclic permutation. Thus Eqs. (27y) and (27z) may be derived from Eq. (27x) by substituting y for x, z for y, and x for z in every instance in which these symbols occur. This method applies fully as well to such expressions as the components of ω as to the general equations of Euler, and will assist the reader in comprehending the otherwise complex nature of this system.

So long as the vector of rotation ω has a finite value, the flow is characterized as rotational, the significance of which may be shown by the following operation: If each component of this vector is reduced to zero, the flow will become irrotational, whereupon,

$$\frac{\partial v_z}{\partial y} = \frac{\partial v_y}{\partial z}; \quad \frac{\partial v_x}{\partial z} = \frac{\partial v_z}{\partial x}; \quad \frac{\partial v_y}{\partial x} = \frac{\partial v_x}{\partial y} \quad (28)$$

Adding and subtracting equivalent terms in Eq. (27x), now written for irrotational motion,

$$\frac{\partial v_x}{\partial t} + v_x \frac{\partial v_x}{\partial x} + \frac{1}{2}v_y \frac{\partial v_y}{\partial x} + \frac{1}{2}v_y \frac{\partial v_x}{\partial y} + \frac{1}{2}v_z \frac{\partial v_x}{\partial z} + \frac{1}{2}v_z \frac{\partial v_z}{\partial x}$$

$$+ \frac{1}{2}v_y \frac{\partial v_y}{\partial x} - \frac{1}{2}v_y \frac{\partial v_x}{\partial y} - \frac{1}{2}v_z \frac{\partial v_x}{\partial z} + \frac{1}{2}v_z \frac{\partial v_z}{\partial x}$$

$$= -\frac{1}{\rho}\frac{\partial}{\partial x}(p + \gamma h)$$

whence

$$\frac{\partial v_x}{\partial t} + v_x \frac{\partial v_x}{\partial x} + v_y \frac{\partial v_y}{\partial x} + v_z \frac{\partial v_z}{\partial x} = -\frac{1}{\rho}\frac{\partial}{\partial x}(p + \gamma h)$$

This result may also be written in the form

$$\frac{\partial v_x}{\partial t} + \frac{\partial (v_x{}^2/2)}{\partial x} + \frac{\partial (v_y{}^2/2)}{\partial x} + \frac{\partial (v_z{}^2/2)}{\partial x} = -\frac{1}{\rho}\frac{\partial}{\partial x}(p + \gamma h)$$

in which the convective portion of the left side of the equation is simply the derivative of the square of the resultant velocity, since

$$\frac{v_x{}^2 + v_y{}^2 + v_z{}^2}{2} = \frac{v^2}{2}.$$

Similar results will be obtained for the two other coordinate directions through the same operation, from which it is at once obvious that throughout the entire flow

$$\frac{\partial v_x}{\partial t} + \frac{\partial E_m}{\partial x} = \frac{\partial v_y}{\partial t} + \frac{\partial E_m}{\partial y} = \frac{\partial v_z}{\partial t} + \frac{\partial E_m}{\partial z} = 0 \qquad (29)$$

The significance of the above result cannot be overemphasized; expressed in words, this important fact is as follows: Except for arbitrary change in the hydrostatic load, in irrotational motion the energy of a particle can vary in magnitude from point to point in the flow only if conditions are changing with time. In other words, steady irrotational flow is always one of constant energy. It will be recalled that in developing Eq. (19) the energy was assumed constant from one stream line to the next, so that

$$\frac{\partial v}{\partial n} - \frac{v}{r} = 0 \qquad \text{or} \qquad \frac{\partial v_s}{\partial n} - \frac{\partial v_n}{\partial s} = 0$$

Referred to the natural coordinate system, this quantity is merely the one component (ω_m) of the rotation vector which can be present in two-dimensional motion, and which must be zero if the motion is to be one of constant energy.

Perhaps even more important is the converse of this conclusion: In unsteady, rotational motion the energy of flow can never be constant, for it will vary at a given point with time, and at a given instant with distance in any direction.

Circulation and Vorticity. To visualize clearly the phenomenon of rotation, it is necessary to introduce a new expression, circulation, which is customarily given the symbol Γ (the Greek capital letter gamma). Circulation is defined as the line integral of the tangential velocity component around any closed curve (see Fig. 21):

$$\Gamma = \oint^L v_L \, dL \qquad (30)$$

It must be understood that such a curve in itself is independent of the velocity field and may be given any desired shape, size, and position. Moreover, as shown in Fig. 22, the circulation around any such three-dimensional curve will be equal to the sum of all the circulations around any system of figures into which the surface bordered by the curve is subdivided. This is true because the line integral along the neighboring sides of each pair of subdivisions will be of opposite sign, so that the algebraic

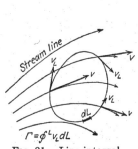
Fig. 21.—Line integral.

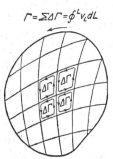
Fig. 22.—Circulation.

sum of all will simply result in the circulation around the outer curve.

If any such surface becomes exceedingly small, the circulation around its enclosing curve will indicate actual rotation about an axis normal to the small increment of area. Thus, the limit of the circulation per unit of area, as the area approaches zero, is equal to twice the component of the rotation vector ω normal to the surface:

$$\lim_{A \to 0} \frac{\Gamma_A}{A} = 2\omega_{\perp A} \qquad (31)$$

The significance of this relationship will be apparent after considering the circulation around the rectangle in the xy plane shown in Fig. 19. If the elementary rectangle is sufficiently small, the line integral of tangential velocity along each side will equal the average velocity between the two corners times the length of the side. Writing the circulation in the counterclockwise direction,

$$\delta\Gamma_{xy} = \left(v_x + \frac{1}{2}\frac{\partial v_x}{\partial x}\delta x\right)\delta x + \left(v_y + \frac{\partial v_y}{\partial x}\delta x + \frac{1}{2}\frac{\partial v_y}{\partial y}\delta y\right)\delta y$$
$$- \left(v_x + \frac{\partial v_x}{\partial y}\delta y + \frac{1}{2}\frac{\partial v_x}{\partial x}\delta x\right)\delta x - \left(v_y + \frac{1}{2}\frac{\partial v_y}{\partial y}\delta y\right)\delta y$$

As the area of the surface approaches the limit zero, the resulting expression for circulation per unit area will be:

$$\lim_{\delta x\delta y\to 0}\frac{\delta\Gamma_{xy}}{\delta x\,\delta y} = 2\omega_z = \frac{\partial v_y}{\partial x} - \frac{\partial v_x}{\partial y}$$

The remaining components of the rotation vector may be derived in a similar manner:

$$\lim_{\delta y\delta z\to 0}\frac{\delta\Gamma_{yz}}{\delta y\,\delta z} = 2\omega_x = \frac{\partial v_z}{\partial y} - \frac{\partial v_y}{\partial z}$$
$$\lim_{\delta z\delta x\to 0}\frac{\delta\Gamma_{zx}}{\delta z\,\delta x} = 2\omega_y = \frac{\partial v_x}{\partial z} - \frac{\partial v_z}{\partial x}$$

Since ω is a vector quantity and similar in many respects to the velocity vector v, its distribution in space may be represented in a similar way; that is, a system of vortex lines may be constructed in space, having the same relation to the rotation or vorticity at any point as a stream line has to velocity. Thus a vortex line shows through tangency to the vorticity vector the sense of rotation and the direction of its axis at every point for a given instant, it being assumed that the vector indicates the direction of rotation of a right-hand screw—*i.e.*, clockwise when looking in the positive direction. If the quantities dx, dy, and dz represent the projections of dw, a differential length of vortex line, on the three axes, Eq. (21) for the stream line will have its counterpart in the differential equation of the vortex line:

$$\frac{dx}{\omega_x} = \frac{dy}{\omega_y} = \frac{dz}{\omega_z} = \frac{dw}{\omega} \tag{32}$$

Just as a small group of stream lines may form a stream filament of varying cross-sectional area ΔA, the rate of discharge past all such sections being the same at a given instant, so may a group of vortex lines form a vortex filament of varying cross-sectional area ΔA (Fig. 23), the circulation around the perimeter of all cross sections being the same at any instant:

$\overset{\centerdot}{v}\,\Delta A = \Delta Q =$ constant   and $2\omega\,\Delta A = \Delta\Gamma =$ constant

Inasmuch as the law of continuity applied to a stream filament thus has its counterpart in the law of constant strength of a

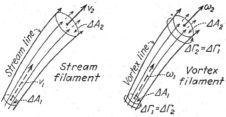

Fig. 23.—Parallel characteristics of stream and vortex filaments.

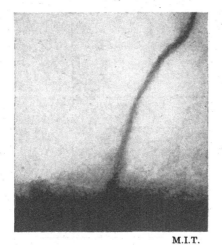

M.I.T.

Fig. 24.—Vortex filament at the base of a weir (compare with Fig. 141).

vortex filament, it might be assumed that the mathematical expression for the divergence of the velocity vector could be written in a similar fashion for the vector of rotation:

$$\operatorname{div} v = \frac{\partial v_x}{\partial x} + \frac{\partial v_y}{\partial y} + \frac{\partial v_z}{\partial z} = 0; \quad \operatorname{div} \omega = \frac{\partial \omega_x}{\partial x} + \frac{\partial \omega_y}{\partial y} + \frac{\partial \omega_z}{\partial z} = 0$$

That this is true may be proved by substituting in the above expression the components of ω in terms of the space derivatives of the velocity.

Characteristics of the Vortex. While stream lines generally exist throughout all portions of a fluid in motion, a single vortex line may exist in an otherwise irrotational flow, so that only those infinitesimal fluid particles lying directly upon that line will undergo rotational motion. Such a case may be illustrated by the movement of fluid in concentric layers around a vertical axis in such a way that all particles not lying upon that

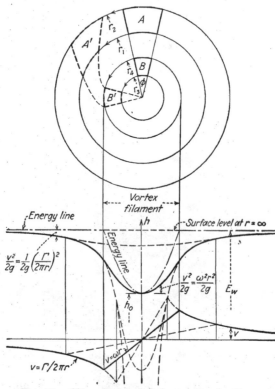

Fig. 25.—Characteristics of the combined vortex.

axis have the same total energy of flow—*i.e.*, the flow is steady and irrotational. If these conditions are fulfilled, the circulation around any horizontal curve (not including the center) must be equal to zero.

Referring to Fig. 25, the circulation around the area A bounded by any two radial lines and any two circular arcs having the radii r_1 and r_2 may be written as follows:

$$\Gamma = v_2 \phi r_2 - v_1 \phi r_1$$

If this quantity is equal to zero, the flow is irrotational and the velocity distribution must be such that

$$v_1\, r_1 = v_2\, r_2 \quad \text{and} \quad v = \frac{C}{r}$$

This is in agreement with Eq. (19) for irrotational flow:

$$\frac{\partial v}{\partial n} = \frac{v}{r}, \quad dn = -dr; \quad \ln v = -\ln r; \quad vr = C$$

Since the total head is constant, the pressure head along any horizontal plane will be

$$\frac{p}{\gamma} = E_w - h - \frac{v^2}{2g} = E_w - h - \frac{C^2}{2gr^2}$$

If the fluid is a liquid with a free surface at constant (atmospheric) pressure, the elevation of this surface above any geodetic datum may be written

$$h = E_w - \frac{C^2}{2gr^2}$$

in which E_w represents the surface elevation an infinite distance from the axis. The profile of such concentric motion is shown in Fig. 25, the phenomenon being known as the free or potential vortex.

Around any circular stream line, however, the circulation will have a finite magnitude

$$\Gamma = 2\pi r v$$

which will be a constant irrespective of the radius of the circle. The term $\Gamma/2\pi$ is then the constant (*i.e.*, the product of the velocity and the radius of curvature) in the above equations. But since the circulation around any such circle, no matter how small, is a finite constant, then the ratio Γ/A becomes infinitely great as the area of the circle approaches zero. Hence, the rotation vector ω is infinite along the axis (the axis being a vortex line) but equal to zero at every other point in the flow.

Obviously, this is an impossible condition, for it requires either that the surface at the center drop an infinite distance below any geodetic datum chosen, or, if the fluid is completely enclosed, that the pressure intensity at the center be negative infinity. However, the space in the neighborhood of the single vortex line may contain fluid which does not follow the potential relationship of the general flow, but behaves as a vortex filament, all particles of which have a constant rate of rotation.

The velocity characteristics of such a filament are similar to those of the tank of liquid rotating with constant angular velocity ω as studied in hydraulics. Since the fluid particles have no motion relative to one another, the velocity must vary directly with the radius:

$$v = \omega r$$

The circulation around any stream line will then vary with the square of the radius

$$\Gamma = 2\pi r v = 2\pi \omega r^2$$

as will the circular area πr^2 bounded by the stream line. Hence, the rotation vector at the center must be

$$\omega = \frac{\Gamma}{2A} = \frac{2\pi \omega r^2}{2\pi r^2} = \omega$$

But the circulation around any area B (Fig. 25) bounded by two radial lines and two circular arcs and not including the central axis, divided by the area of this surface, yields the identical rotation vector for any fluid particle in the vortex filament:

$$\omega = \frac{\Gamma}{2A} = \frac{v_4 \, \phi \, r_4 - v_3 \, \phi \, r_3}{\phi \, (r_4{}^2 - r_3{}^2)} = \omega$$

It is obvious that the energy of flow cannot be a constant, because of this rotational motion. Denoting by h_0 the surface level at the axis, the elevation of the free surface at any radius may be found from Eq. (6n):

$$\frac{v^2}{r} = -\frac{1}{\rho} \frac{\partial}{\partial n}(p + \gamma h) = g \frac{\partial h}{\partial r}$$

$$\frac{v^2}{gr} = \frac{\omega^2 \, r}{g} = \frac{\partial h}{\partial r}; \qquad \int \frac{\omega^2 \, r \, dr}{g} = \int dh$$

$$h = C + \frac{\omega^2 \, r^2}{2g} = h_0 + \frac{v^2}{2g}$$

A combination of the two types of motion is known as the Rankine combined vortex, with profile and velocity distribution as shown in Fig. 25.

Observation with time of the two elementary areas A and B in the light of the foregoing study of the four essential types of displacement should serve to clarify the general conclusions reached at that time. Since a velocity vector exists at every point of the motion with the sole exception of the central axis, all particles are being translated through space at a rate varying with the distance from the central axis. While a mathematical treatment of deformation and rotation in this case requires use of the cylindrical coordinate system, inspection of Fig. 25 will suffice to show that in addition to translation surface A will undergo both linear and angular deformation, but not rotation, whereas surface B will be rotated but not deformed.

Such an isolated vortex filament in an otherwise irrotational state of fluid motion must always be surrounded by a velocity field, as shown in Fig. 25, which extends outward to the boundaries of the flow, the velocity in this field being inversely proportional to the distance from the axis of the vortex filament. The existence of two or more neighboring filaments thus results in a relative movement of each filament in accord with the velocity fields of the others. Two vortices of equal strength and opposite directions of rotation will propel each other in a direction normal to the plane of their axes, as shown in Fig. 26, the velocity of translation depending upon the strength and spacing of the filaments (refer to the curve of velocity distribution as a function of distance from the filament in Fig. 25); on the other hand, if the vortices are of unequal strength, they will move with different velocities around concentric circles of unequal radii. If the filaments are of equal strength and have the same direction of rotation, both will move around the same circle; if of unequal strength, they will travel about concentric circles of unequal radii, on opposite sides of the common center. If a single filament exists near a plane boundary parallel to the axis of the filament, it will move parallel to the boundary as though impelled by an imaginary vortex (its "mirror image") on the other side of the wall. The vortex ring (exemplified by the common smoke ring) moves through space under the influence of its own velocity field; the direction of translation is normal to the plane of the ring. It should be obvious that the presence of many vortices of different strengths in a moving fluid will give rise to a very complex velocity field throughout the fluid.

The reader may readily convince himself of the existence of such movements by generating small vortices with a paddle in a basin of water, the upper ends of the filaments being visible as small depressions in the water surface. The potential vortex is

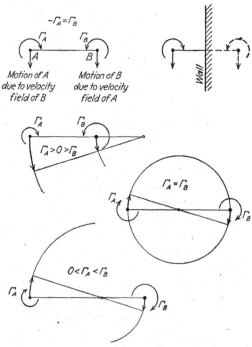

Fig. 26.—Interaction of neighboring vortices.

illustrated by the movement developing in a tank of water shortly after a drain has been opened in the bottom of the tank, a hollow space forming in the surface and extending downward to the drain, and displaying a profile very similar to the outer curve shown in Fig. 25. If the drain is then partially closed, the profile will become that of the Rankine combined vortex. Cyclonic twisters and water spouts are also examples of vortex motion, in which the filament itself is made visible by dust and by water, respectively.

A principle of classical hydrodynamics, known as Thomson's law, states that irrotational motion can never become rotational so long as only gravitational and pressure forces act upon the fluid particles; similarly, under these conditions rotational motion must always remain rotational. Moreover, since the strength of a vortex filament must remain constant throughout

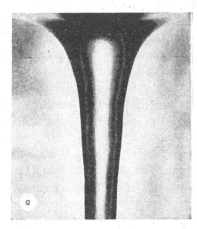

M.I.T.

FIG. 27.—Profiles of (*a*) the potential and (*b*) the combined vortex, occurring above an orifice in the bottom of a tank. The central portion of the combined vortex has been colored by dye.

M.I.T.

FIG. 28.—Velocity distribution of the combined vortex as indicated by the paths of aluminum particles during a short time exposure.

its length, it is impossible for a filament to end abruptly within a fluid medium; it must either be a closed curve, as in the case of a smoke ring, or terminate at a free surface or a solid boundary. While these principles have long since been vindicated by rigid mathematical proof, the fact remains that rotational movement may be started in any fluid at will, the motion then gradually becoming irrotational as the fluid again comes to rest.

These circumstances lead at once to the following conclusion: Since neither weight nor the normal force of fluid pressure upon a particle can produce rotation, any change in rotational energy must be the result of tangential stress, which can be caused only through viscous shear. Indeed, the very basis of modern progress in fluid mechanics lies in the appreciation of the role played by viscosity; the fact has been verified repeatedly in experimental investigations that viscous action is essential to the generation of rotational motion, and that only by conversion into heat through viscous shear can rotational energy be diminished.

It has been shown in the foregoing pages that rotational motion may often be studied quite satisfactorily at any instant in the light of the non-viscous flow equations, regardless of whether the rotation is continuously distributed throughout the fluid, or discontinuously, as in the case of isolated vortex filaments. On the other hand, the rotational characteristics of many types of flow are of so secondary a nature that the assumption of irrotational motion is often fully justified. A more intensive study of rotation will, therefore, be left until a later chapter on viscous flow.

IRROTATIONAL MOTION

The Velocity Potential. Returning now to the case of irrotational movement, since at every point in the field of motion all three components of the vorticity vector ω must equal zero, the following equalities will obtain:

$$\frac{\partial v_z}{\partial y} = \frac{\partial v_y}{\partial z}; \qquad \frac{\partial v_x}{\partial z} = \frac{\partial v_z}{\partial x}; \qquad \frac{\partial v_y}{\partial x} = \frac{\partial v_x}{\partial y} \tag{33}$$

Under these conditions there will exist a velocity potential ϕ (phi) throughout the flow, the space derivatives of which at any point will equal the velocity components in the corresponding directions:

$$v_x = \frac{\partial \phi}{\partial x}; \qquad v_y = \frac{\partial \phi}{\partial y}; \qquad v_z = \frac{\partial \phi}{\partial z} \tag{34}$$

This relationship may be verified by substituting in Eqs. (33) the velocity components as expressed in Eqs. (34):

$$\frac{\partial^2 \phi}{\partial z\, \partial y} = \frac{\partial^2 \phi}{\partial y\, \partial z}; \qquad \frac{\partial^2 \phi}{\partial x\, \partial z} = \frac{\partial^2 \phi}{\partial z\, \partial x}; \qquad \frac{\partial^2 \phi}{\partial y\, \partial x} = \frac{\partial^2 \phi}{\partial x\, \partial y}$$

Since differentiation with respect to two variables is independent of the order of differentiation, the above values will be seen to be identities, and Eqs. (34) are thereby substantiated.

The significance of such a velocity potential may be seen through comparison with the force potential, a mathematical relationship already used in the development and application of the three equations of Euler. It was shown that the derivative of the quantity $-(p + \gamma h)$ in any direction must equal the component of the accelerative force per unit volume in that direction; $-(p + \gamma h)$ is obviously a scalar quantity, its magnitude varying generally with time and three-dimensional space. Thus at any instant imaginary surfaces of constant force potential may be considered to exist throughout the moving fluid, the resultant vector f of accelerative force per unit volume at every point being normal to one of these surfaces. ϕ is likewise a scalar quantity and a function of time and space, and at any instant surfaces of constant velocity potential may also be imagined to exist throughout the fluid, the velocity vector at every point being normal to such a surface. A surface of constant velocity potential is thus normal at all points to the direction of motion, whereas a surface of constant force potential is normal at all points to the direction of acceleration.

Inasmuch as every velocity vector is tangent to a stream line at a given instant, a surface of constant velocity potential must cut all instantaneous stream lines at right angles. Therefore,

$$v = \frac{\partial \phi}{\partial s} \tag{35}$$

and it is apparent that the magnitude of ϕ must increase in the direction of flow according to this relationship. It should also be apparent to the reader that the normal trajectories of the two-dimensional flow net are merely intersections of surfaces of constant velocity potential with the plane of motion. The systematic spacing of these n trajectories resulted in constant increments of ϕ from one line to the next, so that the relative velocity at any point could be seen from the spacing of these potential lines, according to Eq. (35) written in approximate form:

$$v = \frac{\Delta \phi}{\Delta s} = \frac{C}{\Delta s}$$

It has already been shown that flow with constant specific energy must be both irrotational and steady. Even if irrotational motion is unsteady, there is still a direct relationship between the energy of flow and the rate of change of the velocity potential with time, $\partial \phi / \partial t$, which may be developed as follows: Any one of the three equations of Euler written for unsteady irrotational motion (see page 71),

$$\frac{\partial v_x}{\partial t} + v_x \frac{\partial v_x}{\partial x} + v_y \frac{\partial v_y}{\partial x} + v_z \frac{\partial v_z}{\partial x} = -\frac{1}{\rho} \frac{\partial}{\partial x} (p + \gamma h)$$

may now be written in the form,

$$\frac{\partial^2 \phi}{\partial x \, \partial t} + \frac{\partial (v^2/2)}{\partial x} = -\frac{1}{\rho} \frac{\partial}{\partial x} (p_d + p_s + \gamma h)$$

Integrating with respect to x, this will then become:

$$\rho \frac{\partial \phi}{\partial t} + \rho \frac{v^2}{2} + p_d = F(t) - (p_s + \gamma h) \tag{36}$$

The term $F(t)$ represents merely an arbitrary function of time allowing for possible variation of the hydrostatic pressure load on a closed system; it has already been shown that this will have no effect whatsoever upon the pattern of flow (so long as the absolute pressure does not reach zero at any point). Equation (36) may thus be rewritten in the form[1]

$$\rho \frac{\partial \phi}{\partial t} + \rho \frac{v^2}{2} + p_d = \text{constant} \tag{37}$$

showing that in unsteady irrotational flow the dynamic characteristics are directly dependent upon the temporal variation of the velocity potential.

As yet no restriction has been placed on the variation of ϕ, for it remains to be stated that the motion described by the velocity potential must also fulfill the conditions of continuity. The equation of continuity may be written in terms of ϕ, through substitution of its space derivatives [Eqs. (34)] in Eq. (23):

$$\frac{\partial^2 \phi}{\partial x^2} + \frac{\partial^2 \phi}{\partial y^2} + \frac{\partial^2 \phi}{\partial z^2} = 0 \tag{38}$$

This general relationship, known as the equation of Laplace, must be satisfied by any velocity potential.

If, for a given set of boundary conditions, ϕ could be expressed mathematically as a continuous function of time and space which would satisfy Eq. (38), the state of flow would then be completely defined by this one expression; it would at once be possible to determine from it any desired characteristic of flow at any point, through substitution in Eqs. (34) and (37), and thus entirely eliminate further use of the three equations of Euler and the one of continuity—obviously a great simplification.

However, unless a given state of unsteady motion can be reduced to a steady one through introduction of a moving coordinate system, except in certain special cases it will be found exceedingly difficult to express the variation of ϕ with both time and space. Hence, the question of unsteady motion must be

[1] Compare with LAMB, "Hydrodynamics," pp. 19–22.

left for the time being, the discussion of potential flow henceforth dealing entirely with movement that is independent of time. Under these circumstances, the local acceleration will be zero at every point, and Eq. (36) will then become the equation of Bernoulli.

Problems in Three-dimensional Flow. It might now be assumed that ϕ could be written at once as a function of coordinate space for any given boundary conditions. Unfortunately, this is by no means the case. While a velocity potential unquestionably exists for every possible type of irrotational motion, the mathematical ingenuity of the scientist is not yet of such calibre as to enable him to derive at will expressions for more than the simplest cases of such motion. As a matter of fact, classical hydrodynamics advanced largely through the reverse procedure of finding boundary conditions to which known functions of the velocity potential would apply. These known functions for three-dimensional motion are comparatively few in number and deal generally with flow around such bodies of revolution as airships and undersea craft. Since each has its counterpart in the more highly developed study of two-dimensional potential flow, which will be discussed directly, only brief mention of several typical examples will be necessary at this point.

Just as rotation may exist along a single vortex line in otherwise irrotational fluid motion—a singular line, mathematically speaking—it is also permissible to introduce what is known in hydrodynamics as a point source or a point sink—a singular point in the fluid medium at which fluid matter is either created or destroyed at a given constant rate. The equation of continuity for a fluid of constant density will then hold exactly at every point in the flow, with the one exception of the point source or sink, at which point the divergence of the velocity vector will change abruptly from zero to negative or positive infinity. Similar conditions were found to hold in the case of the vortex line, the rotation vector being infinitely great on the line itself, and zero throughout the remainder of the flow.

Since the equation of continuity still applies to all regions of flow other than the one singular point, the rate of discharge Q through all imaginary spheres surrounding the singular point must be the same as the rate at which fluid is presumed to be created or destroyed at that point. Denoting by R the radius of any sphere concentric with the source or sink, since the flow is radial in all directions, the velocity $v = v_R$ at a given radius may be found from the expression

$$v_R = \pm \frac{Q}{4\pi R^2}$$

the plus sign applying to the source and the minus to the sink (Fig. 29). Since this velocity vector is equal to the gradient of the velocity potential in the radial direction, ϕ must then have the following form:

$$\phi = \int v_R \, dR = \mp \frac{Q}{4\pi R}$$

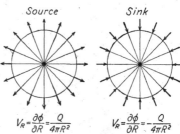

FIG. 29.—Source and sink.

Motion of this sort may be more clearly visualized if compared with flow toward a small orifice in the side of an extremely large tank. With the exception of the immediate neighborhood of the orifice itself, where the motion is curvilinear, the velocity will be radial and will vary inversely with the square of the radius. The orifice symbolizes the point sink, the flow picture in this case being just one-half of the symmetrical pattern of stream lines approaching a true sink from all directions (see Fig. 30). If a source and a sink of equal magnitude are located a distance m apart in a fluid otherwise at rest, all of the fluid leaving the source will return sooner or later to the sink, following stream lines shown schematically in Fig. 31. Placing the cylindrical coordinate origin midway between the source (Q_1) and the sink $(-Q_2)$, the velocity potential of the combined flow will be simply the sum of the velocity potential of the source ϕ_1 and that of the sink ϕ_2; since the radius R of the spherical system now becomes $\sqrt{r^2 + z^2}$ (refer to Fig. 31),

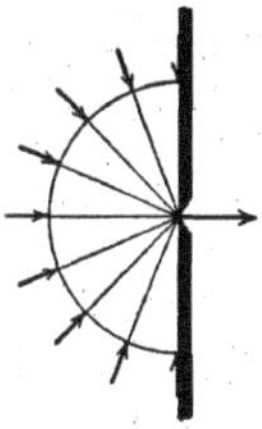

FIG. 30. Flow toward a small orifice.

$$\phi = \phi_1 + \phi_2 = -\frac{Q}{4\pi\sqrt{r^2 + \left(z + \frac{m}{2}\right)^2}} + \frac{Q}{4\pi\sqrt{r^2 + \left(z - \frac{m}{2}\right)^2}}$$

The velocity at any point of the flow will then be the vector sum of the components in the r and z directions, each of which may be found by taking the proper space derivative of ϕ. If the source is greater or smaller than the sink, there will be a positive or negative surplus of fluid $(Q_1 - Q_2)$, which must either go out to, or come in from, infinity, as the case may be.

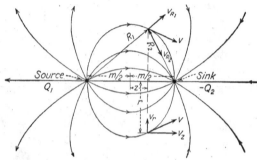

FIG. 31.—Flow between a source and sink of equal strength.

When a source and a sink of equal magnitude are made to approach each other in such manner that the product of rate of discharge and distance Qm is held constant, as m approaches the limit zero the flow pattern will become that of a doublet, shown in Fig. 32. The velocity potential of the doublet (the mathematical development is omitted)[1] will then be

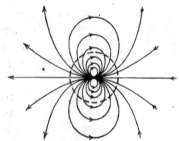

FIG 32.—Three-dimensional doublet.

$$\phi = \frac{Cz}{(r^2 + z^2)^{3/2}}$$

in which C is an arbitrary constant. The velocity at any point may be found in the usual manner.

If there be added to any of these source-sink combinations a linear flow in a direction parallel to the line connecting the singular points, it is evident that the interaction of the two

---

[1] For this derivation, as well as for a more extensive treatment of examples of potential motion, the reader is referred to PRANDTL-TIETJENS' "Fundamentals of Hydro- and Aeromechanics," Engineering Societies Monograph, McGraw-Hill Book Company, Inc., 1934, and to vol. I of W. F. DURAND'S "Aerodynamic Theory," Julius Springer, Berlin, 1934.

systems must produce an entirely new flow pattern. An imaginary surface of revolution may now be considered to separate the flow of the source-sink group from the outer flow, the surface having such a shape that the velocity determined from the combined potential will be tangential to it at all points through which it passes. Since this surface must then be composed of those stream lines forming the boundary between the two systems of flow, the inner flow could be replaced by a solid body of exactly the same surface form without disturbing the flow around it.

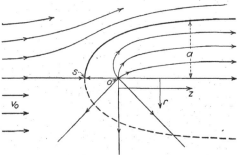

FIG. 33.—Combination of parallel flow and a source—the half body.

As a simple example, consider the combination of a single source with parallel flow. The velocity potential of the linear motion (in the positive z direction) must be:

$$\phi_1 = v_0 z$$

The velocity potential of the source has already been developed. The velocity potential of the resulting flow will be simply the sum of these two:

$$\phi = \phi_1 + \phi_2 = v_0 z - \frac{Q}{4\pi \sqrt{r^2 + z^2}}$$

In Fig. 33 is shown the pattern of stream lines corresponding to this velocity potential. It is obvious that the space occupied by the fluid emerging from the source could be replaced by a solid body of the same form without changing the flow around it in any way, relationships then being at hand for finding the pressure and velocity distribution at any point in the surrounding fluid. Such a profile is known as a half body or a semi-infinite body, since it extends to infinity in the positive z direction.

The velocity component parallel to the z axis may be found by taking the derivative of ϕ in this direction:

$$v_z = \frac{\partial \phi}{\partial z} = v_0 + \frac{Qz}{4\pi(r^2 + z^2)^{3/2}}$$

It may be seen by inspection that the velocity will become parallel and equal to the original magnitude of the oncoming flow only an infinite distance along the z axis in either direction. The position of the nose of the body, which is a point of stagnation, may be determined by placing r and v_z equal to zero in the above expression, with the result:

$$z_s = -\sqrt{\frac{Q}{4\pi v_0}}$$

The imaginary flow through any normal section of the half body to the right of the coordinate origin (refer to **Fig. 33**) must be equal to the discharge from the source,

$$\int^a v_z \, 2\pi r \, dr = Q$$

whereas between the origin and the nose of the body this must equal zero; *i.e.*, all fluid leaving the source in the negative z direction must eventually return and pass to the right. By substituting the value for v_z in this integral expression, an equation may be derived for the radius a of the profile at any given value of z, by which means the exact form of the body may be determined.

Had the velocity potential for a parallel flow been added in similar fashion to that for a source and a sink of equal magnitude, equations would have resulted for flow around a symmetrical body of approximately ellipsoidal form, since under these conditions the discharge from the source would have been entirely absorbed by the near-by sink; the relative dimensions of such a body depend upon the assumed velocity of the oncoming flow and the magnitude and spacing of the source and sink.

The combination of a parallel flow with a doublet results in the pattern of flow around a sphere (**Fig. 34**); the velocity potential of such a flow will be, in terms of the arbitrary doublet constant C,

$$\phi = v_0 z + \frac{Cz}{(r^2 + z^2)^{3/2}}$$

the derivative of which in the z direction yields the axial component of velocity at any point:

$$v_z = v_0 + \frac{C}{(r^2 + z^2)^{3/2}} - \frac{3Cz^2}{(r^2 + z^2)^{5/2}}$$

The magnitude of the doublet constant C may now be determined by writing the conditions at the stagnation point ($r = 0$, $v_z = 0$) in terms of the radius a of the sphere:

$$0 = v_0 + \frac{C}{a^3} - \frac{3C}{a^3}; \qquad C = \frac{v_0 a^3}{2}$$

Introducing this value in the above relationships, the velocity potential and the two velocity components for flow around a sphere of radius a will then be:

$$\phi = v_0 z + \frac{a^3 v_0 z}{2(r^2 + z^2)^{3/2}}$$

$$v_z = v_0 + \frac{a^3 v_0}{2(r^2 + z^2)^{3/2}} - \frac{3}{2}\frac{a^3 v_0 z^2}{(r^2 + z^2)^{5/2}}$$

$$v_r = -\frac{3}{2}\frac{a^3 v_0 z r}{(r^2 + z^2)^{5/2}}$$

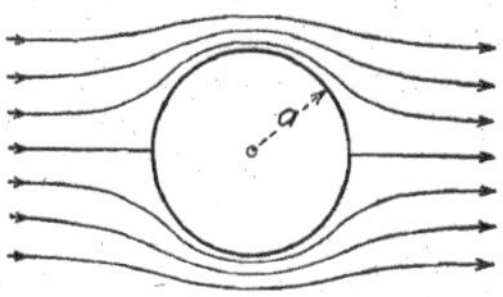

Fig. 34.—Pattern of steady flow around a sphere.

The velocity $v = \sqrt{v_r^2 + v_z^2}$ at any point on the surface of the sphere may now be found through substitution of the term $a = \sqrt{r^2 + z^2}$ in the above expressions. This tangential velocity will vary from zero at the point of stagnation to a maximum value of $1.5v_0$ around the circumference of the sphere in a plane normal to the longitudinal axis.

The intensity of dynamic pressure at any point on the surface of the sphere—and similarly at any point throughout the flow— may be determined from the following general relationship for all such types of flow,

$$\rho\frac{v^2}{2} + p_d = \text{constant} = \rho\frac{v_0^2}{2}; \qquad p_d = \rho\frac{v_0^2 - v^2}{2}$$

since p_d has a magnitude of zero an infinite distance from the sphere, where $v = v_0$. Thus p_d will have a maximum positive value of $\rho v_0^2/2$ at the points of stagnation, and a minimum value of $-\frac{5}{4}\rho v_0^2/2$ in the region of highest velocity.

In each of these cases, the pattern of steady flow around an immersed body has been obtained by superposing parallel motion upon some existing state of movement. Assume now that the body and the surrounding flow are given a rate of motion equal and opposite to the velocity of the parallel flow (the coordinate system thereby remaining stationary); the stream-line pattern of the original flow will evidently be restored to its earlier form—through subtraction of the quantity that had just been added—with the exception that these stream lines now represent the instantaneous pattern for unsteady flow caused by movement of the immersed boundary, as indicated for the sphere and half body in Figs. 32 and 33.

Since the velocity of the parallel flow and the discharge and location of the sources and sinks are entirely optional in any of these problems, it should also be possible to combine as many sources and sinks of different magnitudes as one might desire,

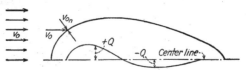

FIG. 35.—Development of the velocity potential for a boundary of complex form.

in order to produce velocity functions for bodies of revolution covering a wide range of profiles. Indeed, this method has been used satisfactorily to develop the velocity potential for flow around an airship—in both forward and lateral motion—by combining with a parallel flow a series of sources and sinks systematically distributed along the axis of the body (see Fig. 35). For the given airship profile and the given velocity of the oncoming flow, the sources and sinks are so grouped that at every point on the profile the normal velocity component caused by the parallel flow is exactly equal and opposite to that caused at the same point by all the sources and sinks together. That is, the resulting normal velocity component must be zero over the entire surface. Obviously, the total discharge from the sources must equal the total discharge into the sinks. When these conditions are satisfied, the resulting velocity potential may be determined through summation.

By carefully grouping sources and sinks in this manner, and by addition of ring vortices in planes normal to the axis, it is possible to determine the velocity and pressure distribution around a body of revolution of almost any desired form, although the computations will become quite involved. If carried to an extreme, this method might be expected to yield results for bodies of any irregular shape, through groups of sources, sinks, and ring vortices at other points than the axis; the complicated form of the resulting velocity potential, however, would hardly warrant such a procedure.

Two-dimensional Flow: the Stream Function. If the conditions of steady, irrotational flow are such that the motion is entirely two-dimensional—that is, if the stream-line pattern is identical in a series of parallel planes (xy), there being no movement whatever in a direction (z) normal to these planes—the relationships just developed for three-dimensional potential flow become greatly simplified. The principle of continuity [Eq. (23)], which must hold for any type of flow at constant density, reduces to

$$\operatorname{div} \boldsymbol{v} = \frac{\partial v_x}{\partial x} + \frac{\partial v_y}{\partial y} = 0 \tag{39}$$

whereas the fact that in two-dimensional motion only one component (ω_z) of the vorticity vector could possibly exist, leads to the following simplification of Eqs. (33) for irrotational motion:

$$\frac{\partial v_y}{\partial x} = \frac{\partial v_x}{\partial y} \tag{40}$$

The Bernoulli theorem, of course, requires no modification, while Eqs. (34) will become,

$$v_x = \frac{\partial \phi}{\partial x}; \qquad v_y = \frac{\partial \phi}{\partial y} \tag{41}$$

Substitution of these components in the equation of continuity results in the equation of Laplace for two-dimensional flow, a relationship that must be satisfied by any true velocity potential:

$$\frac{\partial^2 \phi}{\partial x^2} + \frac{\partial^2 \phi}{\partial y^2} = 0 \tag{42}$$

Moreover, because of the condition expressed by Eq. (39), there will also exist a stream function ψ (psi), the derivatives of which in the two coordinate directions yield the velocity components in directions normal to the respective axes:

$$v_x = \frac{\partial \psi}{\partial y}; \qquad v_y = -\frac{\partial \psi}{\partial x} \tag{43}$$

Since the velocity vector at any point is tangent to the stream line passing through that point, it is evident that the components of v must be proportional to the respective components of an increment of distance ds along the stream line. Thus, the differential equation of the stream line must be as follows:

$$\frac{dx}{v_x} = \frac{dy}{v_y} \quad \text{or} \quad v_x\,dy - v_y\,dx = 0 \tag{44}$$

Introducing the components of the velocity vector expressed as derivatives of ψ into this expression,

$$\frac{\partial \psi}{\partial y}\,dy + \frac{\partial \psi}{\partial x}\,dx = d\psi = 0 \tag{45}$$

from which it will be seen that the stream line is that line along which the stream function is constant. Substitution of Eq. (43) in Eq. (40),

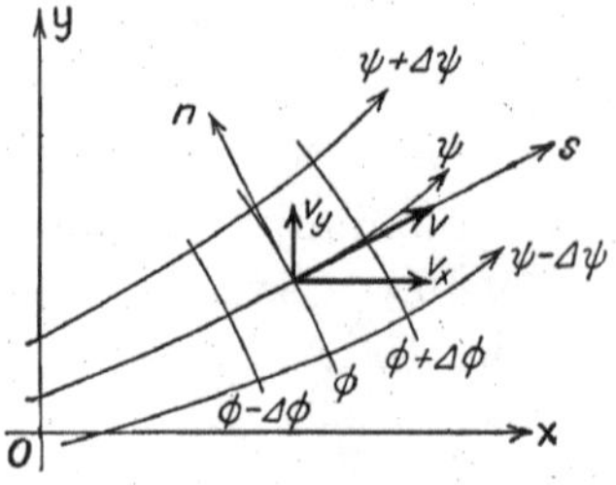

$$\frac{\partial^2 \psi}{\partial x^2} + \frac{\partial^2 \psi}{\partial y^2} = 0 \tag{46}$$

shows that the stream function must also satisfy the equation of Laplace. Since each component of the velocity vector is now expressible in terms of either ϕ or ψ, the following identities must exist:

Fig. 36.—Stream and potential lines.

$$\frac{\partial \phi}{\partial x} = \frac{\partial \psi}{\partial y}; \quad \frac{\partial \phi}{\partial y} = -\frac{\partial \psi}{\partial x} \tag{47}$$

That is, lines of constant velocity potential and lines of constant stream function must always intersect each other at right angles. It should now be clear to the reader that the stream lines and normal trajectories of the two-dimensional flow net are really lines of constant ψ and constant ϕ, respectively, so placed that the spacing between every pair of stream lines and every pair of potential lines represents a constant increment of ψ and of ϕ. As will be clear from Fig. 36, according to the natural coordinate notation,

$$\frac{\partial \phi}{\partial s} = v = \frac{\partial \psi}{\partial n} \tag{48}$$

and

$$\frac{\partial \phi}{\partial n} = 0 = \frac{\partial \psi}{\partial s} \tag{49}$$

Hence, since the product $v\,dn$ represents the increment of rate of discharge dq per unit distance normal to the plane of motion (*i.e.*, volume per unit time per unit length equals area per unit time), it will be evident that the quantity of fluid passing per second between any two stream lines Δq must be both dimensionally and numerically equal to the change in magnitude of the stream function $\Delta\psi$ from one stream line to the next:

$$\Delta q = \int^{\Delta n} v\,dn = \Delta\psi \tag{50}$$

This is, of course, merely another way of expressing the law of continuity for two-dimensional irrotational motion.

Just as in the case of three-dimensional potential motion, the velocity potential for certain types of two-dimensional flow may be formulated analytically through combination of sources, sinks, circulation, and parallel motion. Since the flow must be entirely planar in such problems, the stream-line patterns may often be found conveniently through graphical combination of the elementary flow patterns, in the manner indicated early in Chap. II. In many cases, however, the boundary conditions make an exact mathematical derivation of the velocity potential either impossible or at best extremely difficult—despite the fact that a velocity potential must exist for every case of steady, irrotational motion. In such instances, the construction of the flow net by graphical means, without heed to the velocity potential as a mathematical function, is the most practical means of solution. This procedure will, of course, yield the same results for velocity distribution as would differentiation of ϕ, were the latter function known, for the construction of the flow net is based upon identical physical relationships. In addition to the foregoing methods of solution, there exists a very significant type of mathematical operation involving the use of complex variables, whereby the number of available expressions for the velocity potential is tremendously increased. Owing to the nature of this method, it will be discussed separately in the following chapter.

CONFORMAL MAPPING

Theory of Complex Variables. A complex number is one composed of real and imaginary terms. Any imaginary term, such as $\sqrt{-a}$, may be reduced to the product of a real quantity, $\sqrt{a}$, and the unreal or imaginary root $\sqrt{-1}$, the latter commonly being given the symbol i; an imaginary quantity, hence, always contains the factor i. Thus a complex number z may be written as the sum of a real term x and an imaginary term iy,

$$z = x + iy \tag{51}$$

in which the real and imaginary terms must be regarded as distinct from one another.

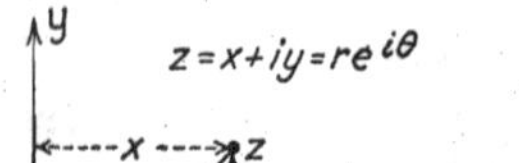

Fig. 37.—Graphical representation of a complex number.

Such a number may be plotted, however, by presuming an imaginary axis along which is measured the real part y of the imaginary term, and a real axis at right angles to it along which is measured the real term x of the complex quantity. Thus in the z plane, according to the above relationship, the point z lies a distance x along the horizontal or real axis, and the distance y along the vertical or imaginary axis, the magnitude of the complex number z then being given by the sum $x + iy$, as shown in Fig. 37.

Polar notation may also be used to express the value of a complex variable. Denoting by r the distance of any point z from the coordinate origin, and by θ the angle between this radius vector and the real axis x, the following relationships will be seen to hold:

$$r = \sqrt{x^2 + y^2}, \quad \text{or} \quad x = r \cos \theta, \quad y = r \sin \theta$$

Hence

$$z = r (\cos \theta + i \sin \theta) \tag{52}$$

The radius r is known as the modulus of z, and the angle θ as the amplitude or argument of z. Refer to Fig. 37.

If the quantity $e^{i\theta}$ is expanded in a series,

$$e^{i\theta} = 1 + \frac{i\,\theta}{1} - \frac{\theta^2}{1\cdot 2} - \frac{i\,\theta^3}{1\cdot 2\cdot 3} + \frac{\theta^4}{1\cdot 2\cdot 3\cdot 4} + \frac{i\,\theta^5}{1\cdot 2\cdot 3\cdot 4\cdot 5}$$

and the real and imaginary terms separated,

$$\left(1 - \frac{\theta^2}{1\cdot 2} + \frac{\theta^4}{1\cdot 2\cdot 3\cdot 4} - \cdots\right) = \cos\theta$$

$$i\left(\frac{\theta}{1} - \frac{\theta^3}{1\cdot 2\cdot 3} + \frac{\theta^5}{1\cdot 2\cdot 3\cdot 4\cdot 5} - \cdots\right) = i\sin\theta$$

it is at once evident that Eq. (52) may be written in the more convenient form:

$$z = re^{i\theta} \quad (53)$$

Inasmuch as a complex number may be represented vectorially by the direction and magnitude of r, it will be clear that the sum or difference of two complex numbers may be obtained by the vector sum or difference of their respective moduli (see Fig. 38).

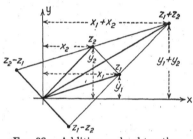

Fig. 38.—Addition and subtraction.

From Eq. (52) it is obvious that the same result may be obtained by adding or subtracting the real and imaginary parts of the two complex quantities, as shown in the illustration; that is,

$$z_3 = z_1 + z_2 = x_1 + i\,y_1 + x_2 + i\,y_2$$
$$= (x_1 + x_2) + i\,(y_1 + y_2)$$

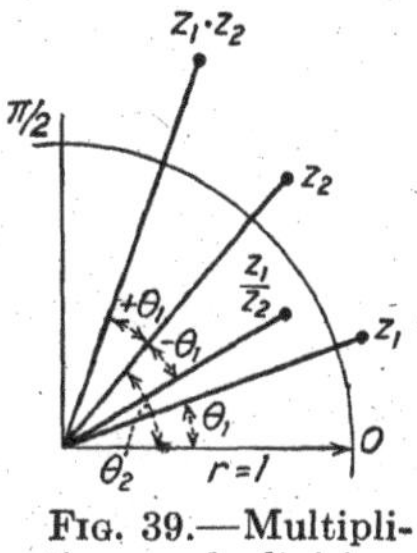

Multiplication or division of one complex number by another may be performed by taking the product or quotient of the moduli and the sum or difference of the arguments,

Fig. 39.—Multiplication and division.

since

$$z_3 = z_1 \cdot z_2 = r_1\,e^{i\theta_1} \cdot r_2\,e^{i\theta_2} = r_1\,r_2\,e^{i(\theta_1+\theta_2)}$$

The operation is shown in Fig. 39. In the same manner, the power of a complex number is found through raising the modulus to the power and multiplying the argument by the exponent, as illustrated in Fig. 40, since

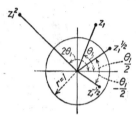

Fig. 40.—Powers of complex numbers.

$$z_2 = z_1{}^n = (r_1\, e^{i\theta_1})^n = r_1{}^n\, e^{in\theta}$$

Evidently, the relative change as a result of the operation depends in part upon the proximity of the original complex quantity to the unit circle, for which $r = 1$.

The logarithm of a complex number may be obtained by a combination of rectangular and polar coordinates. From Eqs. (53) and (51),

$$z_2 = \ln z_1 = \ln r_1\, e^{i\theta_1} = \ln r_1 + i\theta_1 = x_2 + iy_2$$

Therefore,

$$x_2 = \ln r_1; \qquad y_2 = \theta_1$$

As shown in Fig. 41 the effect of this operation upon z will depend upon the magnitude of the modulus with respect to the

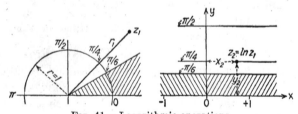

Fig. 41.—Logarithmic operations.

unit circle, and upon its direction with respect to the coordinate axes. That is, if r_1 in the polar notation is greater than unity, x_2 in the rectangular notation will be positive; if r_1 is less than unity, x_2 will be negative; y_2 will always be equal to θ_1 plotted vertically in radians. Thus, the shaded areas in the rectangular plane correspond to the logarithms of the similarly shaded areas in the polar plane.

Significance of Conformal Representation. Performing such operations upon a given complex variable is the essence of the mathematical process known as conformal mapping. If one complex variable can be expressed as a function of another complex variable, and if either one of these can also be expressed as a known function of coordinate space, then the expression of the other as a function of coordinate space may be readily derived.

Assume, now, that one complex variable w is plotted in the w plane, of which the ψ axis is imaginary and the ϕ axis is real, and that w is a function of another complex variable z, which is plotted in the z plane. Then

$$w = \phi + i\psi = f(z) = f(x + iy) \tag{54}$$

Assume further that in the w plane are drawn two families of equidistant lines, parallel, respectively, to the real and imaginary axes, as shown in Fig. 42. These will denote lines of constant ψ and constant ϕ, the increments $\Delta\psi$ and $\Delta\phi$ being equal throughout

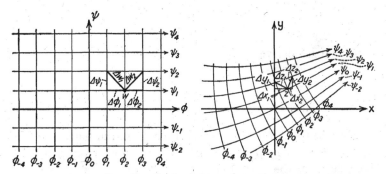

FIG. 42.—Conformal transformation.

the system. If z now be assumed to vary in such manner that the real part (ϕ) of w changes, the imaginary part ($i\psi$) remaining constant, there will be traced in the z plane a line conforming to this line of constant ψ in the w plane. If this procedure is followed for both families of parallel lines in the w plane, there will result a conformal map of this network in the z plane, the shape and position of the transformed image depending upon the functional relationship between w and z as expressed generally in Eq. (54), and shown schematically in Fig. 42.

Certain pertinent characteristics of this operation must be noted. First of all, it is evident that both ϕ and ψ are real functions of x and y, and vice versa:

$$\phi = \phi(x, y), \quad \psi = \psi(x, y); \quad \text{or} \quad x = x(\phi, \psi), \quad y = y(\phi, \psi)$$

That is, for a given functional relationship of the two complex variables, the real quantities x, y, ϕ, and ψ will also be interrelated in a manner independent of the imaginary factor i. Thus, a knowledge of the function will provide a means of reproducing in a real xy plane the transformed $\phi\psi$ network; if the characteristics of the original network are known, then the characteristics of the transformed network will also be known.

Since for every point w_1 in the w plane there is a corresponding point z_1 in the z plane according to the relationship of Eq. (54), it follows that for every change dw in the one complex variable there must be a corresponding change dz in the other. These increments, furthermore, must be equal to the sum of the variations of their real and imaginary parts, respectively; hence, the ratio of the two corresponding increments will be equal to:

$$\frac{dw}{dz} = \frac{d\phi + i\, d\psi}{dx + i\, dy}$$

But the partial derivatives of z with respect to x and to y, [see Eq. (51)]

$$\frac{\partial z}{\partial x} = 1; \qquad \frac{\partial z}{\partial y} = i$$

when substituted in the expressions for the partial derivatives of w with respect to x and to y will yield the following pertinent relationships for the variation of w with respect to z:

$$\frac{\partial w}{\partial x} = \frac{dw}{dz}\frac{\partial z}{\partial x} = \frac{dw}{dz}; \qquad \frac{\partial w}{\partial y} = \frac{dw}{dz}\frac{\partial z}{\partial y} = i\frac{dw}{dz}$$

Therefore,

$$\frac{dw}{dz} = \frac{\partial w}{\partial x} = \frac{\partial \phi}{\partial x} + i\frac{\partial \psi}{\partial x}$$

and

$$\frac{dw}{dz} = \frac{1}{i}\frac{\partial w}{\partial y} = \frac{1}{i}\left(\frac{\partial \phi}{\partial y} + i\frac{\partial \psi}{\partial y}\right)$$

Hence,

$$\frac{dw}{dz} = \frac{\partial \phi}{\partial x} + i\frac{\partial \psi}{\partial x} = \frac{1}{i}\frac{\partial \phi}{\partial y} + \frac{\partial \psi}{\partial y} \tag{55}$$

Since the real and imaginary parts of any complex relationship are quite distinct, it is evident that the real terms in the above equation must equal each other and that the imaginary terms must also be equal:

$$\frac{\partial \phi}{\partial x} = \frac{\partial \psi}{\partial y}; \qquad \frac{\partial \phi}{\partial y} = -\frac{\partial \psi}{\partial x}$$

These two equalities, known as the Cauchy-Riemann equations, are identical with Eqs. (47); since Eq. (54) may be shown to satisfy the equation of Laplace, ϕ and ψ represent the velocity potential and the stream function for two-dimensional irrotational flow; hence,

$$\frac{\partial \phi}{\partial x} = v_x = \frac{\partial \psi}{\partial y}; \qquad \frac{\partial \phi}{\partial y} = v_y = -\frac{\partial \psi}{\partial x} \qquad (56)$$

Introducing these velocity components into Eq. (55),

$$\frac{dw}{dz} = v_x - i \, v_y \qquad (57)$$

from which it is evident that the velocity may also be considered a complex variable of the form $v = v_x - i \, v_y$, which is then a function of $z = x + i \, y$. From this expression may be obtained the magnitude and direction of the velocity vector, although v itself is not a true vector, for it has both real and imaginary parts. Since the two complex variables v and w are functions of a third complex variable z, it follows that a functional relationship must also exist between w and v:

$$w = f(v) = f(v_x - i \, v_y) \qquad (58)$$

It is this general expression which permits solution of certain problems of flow from solid boundaries into the atmosphere, as in the case of jets and weir nappes.

From Eq. (55) it will be obvious that the variation of w with respect to z is independent of the direction of the increment dz—i.e., independent of the ratio dy/dx. In Fig. 42, for instance, as the diagonals Δz_1 and Δz_2 approach zero, the limits of $\Delta w_1/\Delta z_1$ and $\Delta w_2/\Delta z_2$ will be identical, despite the differences between Δx_1 and Δx_2, and Δy_1 and Δy_2. Since the derivative dw/dz depends upon the position, but not the direction, of dz, it thus can represent only the magnitude of rotation and linear distortion involved in the transformation, for from Eqs. (56) it is apparent

that the transformed flow net has the identical angular characteristics of the original—that is, all lines of the net still meet at
right angles, while every "mesh" of the net is still a perfect
square when reduced to the infinitesimal. Thus, the operation
of conformal transformation will change the linear characteristics
of a network of lines without affecting in any way whatever the
angular intersections.

A familiar application of this operation is illustrated by the
Mercator projection used in mapping. In this projection the
parallels of latitude and the great circles of longitude on the terrestrial globe are reproduced in a two-dimensional plane as two
rectilinear families of parallel lines. At the equator the linear
distortion is zero, but the poles are changed from points to lines
of the same length as the equator. Since no angular distortion
results from this operation, such "infinitesimal" areas as the
maps of towns and cities will remain practically unchanged in
shape, despite the obvious distortion of large continental areas
at high latitudes. Moreover, if the same method is used to
show the polar region, the two-dimensional projection will be a
series of radial lines representing the longitudinal meridians, and
a series of concentric circles of constant latitude, of which
one is the equator. Obviously, conditions at the pole remain
unchanged, whereas the regions farthest from the pole undergo
the greatest distortion. Again, however, angular characteristics
are unaffected, and infinitesimal areas that were originally square
will remain so. If now the first projection be called the w plane
and the second the z plane, it is apparent that a single functional
relationship must exist between the two, since every point in the
one has its counterpart in the transformed or conformal map,
and vice versa.

Elementary Transformations. As a preliminary exercise
in the methods of conformal mapping, consider the function

$$w = \phi + i\,\psi = f(z) = f(x + i\,y) = (a + i\,b)\,z$$

At once one may write

$$\phi + i\,\psi = (a + i\,b)(x + i\,y) = ax + ibx + iay - by\,.$$

from which it is apparent that through equating the real and
imaginary parts there will result:

$$\phi = ax - by; \quad \psi = bx + ay$$

The velocity components may be found directly through differentiation of the velocity potential or the stream function, or through the operation:

$$\frac{dw}{dz} = v_x - i v_y = a + i b$$

Again equating the real and imaginary parts,

$$v_x = a; \quad v_y = -b$$

The lines of constant ϕ and constant ψ in the z plane may be found by setting the expressions just found for ϕ and ψ equal to successive numerical constants, the increments always being

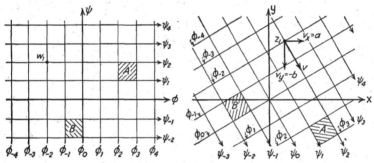

FIG. 43.—Conformal representation of the function $w = (a + i b)z$.

the same. If a and b are both positive, the original and the transformed flow nets will be as shown in Fig. 43, the slope and spacing of the conformal lines depending upon the magnitude of a and of b. If these constants are equal, the slope will be 45°; if a is positive, the horizontal component of flow will be to the right—if negative, to the left; if b is positive, flow will have a downward component—if negative, an upward component; if a is zero, the flow will be vertical; similarly, if b is zero, the flow will be horizontal. Obviously, if b is zero and a is unity, the pictures in the two planes will be identical. Similarly, if a is zero and b is unity, although the scale will be the same, the stream lines will be changed to potential lines, and the potential lines to stream lines. It is thus evident that multiplication of any function by i will rotate the pattern of motion through 90°, while multiplication by $i^2 = -1$ will completely reverse the direction of flow.

A general function of the type

$$w = az^{\frac{\pi}{\alpha}}$$

may through the relationship $z = re^{i\theta}$ be written in the form

$$w = a\,r^{\frac{\pi}{\alpha}} e^{\frac{i\pi\theta}{\alpha}} = ar^{\frac{\pi}{\alpha}}\left(\cos\frac{\pi\theta}{\alpha} + i\sin\frac{\pi\theta}{\alpha}\right)$$

from which it is apparent that the velocity potential and the stream function will be as follows:

$$\phi = a\,r^{\frac{\pi}{\alpha}} \cos\frac{\pi\theta}{\alpha}; \qquad \psi = a\,r^{\frac{\pi}{\alpha}} \sin\frac{\pi\theta}{\alpha}$$

The flow net resulting from this transformation will evidently

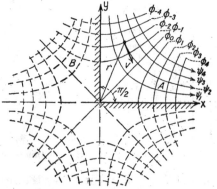

FIG. 44.—Potential flow at a 90° boundary angle.

depend in form upon the magnitude of the angle α. If this angle is 90°,

$$w = a\,r^2\,e^{2i\theta}$$

and

$$\phi = a\,r^2 \cos 2\theta = a\,(x^2 - y^2); \qquad \psi = a\,r^2 \sin 2\theta = 2axy$$

The velocity components may now be found through differentiation of the velocity potential, giving

$$v_x = \frac{\partial \phi}{\partial x} = 2ax; \qquad v_y = \frac{\partial \phi}{\partial y} = -2ay$$

and

$$v = \sqrt{v_x^2 + v_y^2} = 2a\sqrt{x^2 + y^2} = 2ar$$

A plot of the lines of constant stream function, with constant increments between lines, will result in a series of hyperbolas in the four quadrants of the polar coordinate plane. Selecting the axes x and y as solid boundaries of flow, the stream and potential lines in the first quadrant (see Fig. 44) will then form the flow net for potential motion at a 90° angle in the boundary, there being a point of zero velocity at the corner.

Since the angle α evidently represents the angle between the two straight boundary walls, this function may be used to determine the flow characteristics at a corner of any desired angle. Use of the angle $\alpha = \pi$ must then yield parallel flow along a straight wall, as may be seen from the form of the original function.

Source-sink Combinations. The function

$$w = a \ln z$$

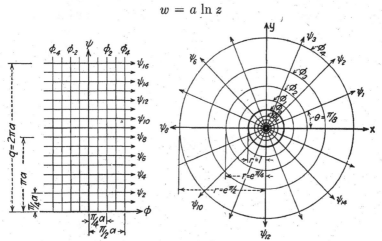

FIG. 45.—Transformation from parallel flow to flow from a source.

may be treated as follows:

$$\phi + i\psi = a \ln re^{i\theta} = a \ln r + ai\theta$$
$$\phi = a \ln r; \qquad \psi = a\theta$$
$$v_r = \frac{\partial \phi}{\partial r} = \frac{a}{r}; \qquad v_t = \frac{\partial \phi}{r\, \partial \theta} = 0$$

Since flow that is entirely radial, the tangential component cf the velocity being zero at all points, must be flow from a two-dimensional source (if a is positive) or into a sink (if a is negative) then the constant a may be expressed in terms of the two-dimensional rate of discharge q:

$$q = 2\pi r v_r = 2\pi a; \qquad a = \frac{q}{2\pi}$$

The transformation is shown in Fig. 45.

The similar function

$$w = a\, i\, \ln z$$

may be treated in a corresponding manner:

$$\phi + i\psi = a\, i\, \ln r e^{i\theta} = a\, i\, \ln r - a\theta$$
$$\phi = -a\theta; \qquad \psi = a \ln r$$
$$v_r = \frac{\partial \phi}{\partial r} = 0; \qquad v_t = \frac{\partial \phi}{r\, \partial \theta} = -\frac{a}{r}$$

Since in this case the velocity is entirely tangential, the transformed image represents flow in concentric circles with constant

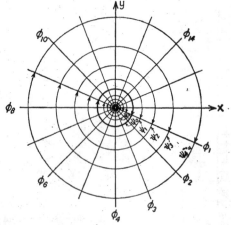

Fig. 46.—Potential flow with constant circulation.

circulation; the constant a will then be (assuming Γ positive in the clockwise direction):

$$-\Gamma = 2\pi r v_t = -2\pi a; \qquad a = \frac{\Gamma}{2\pi}$$

The flow pattern in the z plane may be seen from Fig. 46.

If a composite function be written in the form

$$w = \frac{q + i\Gamma}{2\pi} \ln z$$

the operation will result in a family of logarithmic spirals, as illustrated in Fig. 47, the direction and curvature of the spirals depending upon the sign and magnitude of q and Γ. It is evident that this flow pattern may also be obtained graphically according to the method already discussed of adding velocity fields. As will be seen from the foregoing development, the velocity potential and the stream function of the family of logarithmic spirals are merely the sums of those for radial and concentric motion.

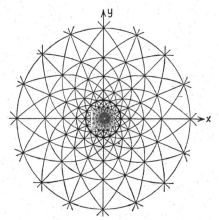

FIG. 47.—Combination of potential vortex with source or sink.

A source and sink of equal magnitude and spaced a distance $2m$ apart will result from the function of transformation

$$w = \frac{q}{2\pi} \ln \frac{z+m}{z-m}$$

Referring to the flow net shown in Fig. 48 (which may also be constructed graphically through combination of the stream and potential lines of the source and sink), it will be seen that

$$w = \phi + i\psi = \frac{q}{2\pi} \ln \frac{r_1 e^{i\theta_1}}{r_2 e^{i\theta_2}} = \frac{q}{2\pi} \left[\ln \frac{r_1}{r_2} + i(\theta_1 - \theta_2) \right]$$

from which the velocity potential and the stream function may be found by inspection:

$$\phi = \frac{q}{2\pi} \ln \frac{r_1}{r_2}; \quad \psi = \frac{q}{2\pi} (\theta_1 - \theta_2)$$

Thus, potential lines and stream lines will all be circles, the loci of $r_1/r_2 =$ constant and $(\theta_1 - \theta_2) =$ constant, respectively.

Through use of Eq. (57) it may be shown that the magnitude of the velocity vector at any point will be

$$v = \frac{m\,q}{\pi\,r_1\,r_2}$$

although the development of this relationship is too involved to be presented in this text. This velocity may also be found by

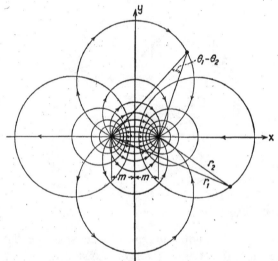

Fig. 48.—Source and sink of equal strength.

adding vectorially the velocities of flow from the source and toward the sink at any point.

As in the foregoing example, the picture of flow from a source to a sink may be transformed into the pattern of circulation around two vortex lines simply by multiplying the original function of z by the factor i. In this way the potential lines in Fig. 48 will become stream lines, and vice versa, and the flow net will represent the unsteady flow pattern of the two vortex filaments in the upper left-hand corner of Fig. 26. The pattern of steady flow may be developed by adding graphically a parallel flow equal and opposite to the velocity of translation of the two filaments; that is, $v_0 = \Gamma/4\pi m$. Obviously, the outline of either filament must be formed by one of the circular stream lines, it being of no consequence whether or not the two filaments are of the same diameter so long as the circulations around their circumferences are of the same magnitude and opposite sign.

Any portion of any potential flow may be replaced by a solid boundary whose outline coincides with a stream line without changing the characteristics of flow in any way. Thus, the foregoing pattern of circulatory motion may be used to find the approximate velocity and pressure distribution for flow under a movable cylindrical weir profile by fitting one stream line of the flow net to the cross section of the cylinder, as shown in Fig. 49.[1] This method is not exact, because the lines of flow upstream from the weir deviate from those of the potential function, although the latter may be used to good advantage as the first approximation in the graphical construction of the net. Moreover, the point at which the flow separates from the cylinder moves upstream with increasing head; that this will affect the velocity of flow will be apparent from the fact that only through knowing the elevation of this point of atmospheric pressure can one establish the velocity corresponding to the given spacing of stream lines.

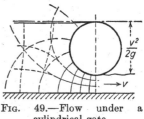

FIG. 49.—Flow under a cylindrical gate.

Flow around a Cylinder. If the distance m in the source-sink combination approach zero, the product mq thereby remaining constant, the flow pattern will approach that of the two-dimensional doublet, or dipol, shown in Fig. 50. The corresponding complex function will then have the form

$$w = \frac{a}{z}$$

The velocity potential and the stream function are found as follows:

$$\phi + i\psi = \frac{a}{x + iy} = \frac{a(x - iy)}{(x + iy)(x - iy)} = \frac{ax}{x^2 + y^2} - \frac{aiy}{x^2 + y^2}$$

$$\phi = \frac{ax}{x^2 + y^2} = \frac{a\cos\theta}{r}; \quad \psi = -\frac{ay}{x^2 + y^2} = -\frac{a\sin\theta}{r}$$

[1] For extensive application of conformal mapping to the design of hydraulic structures, see KULKA, H., "Der Eisenwasserbau," vol. I, Wilhelm Ernst & Son, Berlin, 1928.

Both ϕ and ψ are thus constant along circles passing through the coordinate origin; in Fig. 50 ψ and ϕ are increased successively by constant increments, the radii of the stream and potential lines

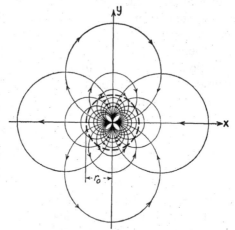

Fig. 50.—Two-dimensional doublet.

varying inversely with these quantities. The velocity components will be found to be

$$v_x = \frac{\partial \phi}{\partial x} = \frac{a\,(y^2 - x^2)}{(x^2 + y^2)^2}; \qquad v_y = \frac{\partial \phi}{\partial y} = \frac{-2axy}{(x^2 + y^2)^2}$$

from which the magnitude of the velocity vector at any point will equal

$$v = \sqrt{v_x^2 + v_y^2} = \sqrt{\frac{a^2\,(y^2 - x^2)^2 + 4a^2 x^2 y^2}{(x^2 + y^2)^4}} = \frac{a}{x^2 + y^2} = \frac{a}{r^2}$$

Comparable to the three-dimensional doublet, from this flow net may be determined the unsteady stream-line picture resulting from the motion through a fluid of an infinitely long circular cylinder, simply by constructing upon the flow net a circle concentric with the coordinate origin to represent the cross section of the cylinder. Designating by v_0 the velocity of translation of the cylinder, the magnitude of a may be found by setting equal to v_0 the velocity of the fluid immediately in front (or in back) of the body (refer to Fig. 50), at the distance r_0 from the origin:

$$v = \frac{a}{r_0^2} = v_0 \qquad \text{and} \qquad a = v_0\,r_0^2$$

Obviously the usual methods cannot be used to find the pressure distribution around the body, for unsteady motion denotes variable energy of flow. The steady picture, however, may be obtained directly through adding (either graphically or analytically) to the above potential and stream functions those of a parallel flow with the velocity v_0—that is, the body is brought to rest through superposing vectorially upon the entire flow picture a velocity v_0 equal and opposite to that of the cylinder:

$$\phi = v_0 \, x + \frac{v_0 \, r_0^2 \, x}{x^2 + y^2}; \qquad \psi = v_0 \, y - \frac{v_0 \, r_0^2 \, y}{x^2 + y^2}$$

This is satisfied by the complex function

$$w = v_0 \left(z + \frac{r_0^2}{z} \right)$$

since

$$w = \phi + i \, \psi = v_0 \, (x + i \, y) + \frac{v_0 \, r_0^2 \, (x - i \, y)}{x^2 + y^2}$$

The velocity components may be found either by differentiating the above relationships or by adding algebraically to v_x for unsteady motion the quantity v_0; v_y, of course, must remain unchanged by this operation. Furthermore, since $r^2 = x^2 + y^2$, $x = r \cos \theta$, and $y = r \sin \theta$, the velocity at all points around the circumference of the cylinder may be found from the relationship:

$$v = v_t = 2v_0 \sin \theta, \qquad \text{when} \qquad r = r_0$$

The pressure distribution around the cylinder may now be found from the expression already developed,

$$p_d = \frac{\rho}{2} \, (v_0^2 - v^2)$$

from which it will be seen that there is a maximum pressure intensity of $\rho\frac{v_0^2}{2}$ at both points of stagnation, and a minimum intensity of $-3\rho\frac{v_0^2}{2}$ at either intersection of the circle with the y axis. The stream-line pattern for flow around a cylinder is shown in Fig. 51; this flow net may be found by combining graphically a parallel flow with a doublet, the spacing of the parallel stream lines for a given doublet depending only upon the diameter of the cylinder; such spacing may then represent any desired magnitude of the parameter $\rho\dfrac{v_0{}^2}{2}$.

If one now superpose upon the flow picture just found the pattern of stream lines for a constant positive circulation Γ around the cylinder, graphical addition of the two systems will yield a flow net resembling that shown in Fig. 52. As can be seen

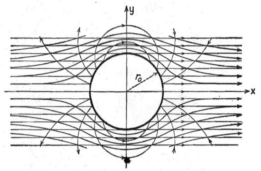

Fig. 51.—Potential flow around a cylinder—graphical combination of parallel flow with a doublet.

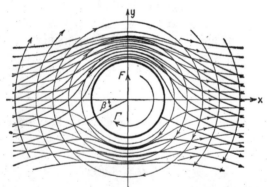

Fig. 52.—Displacement of stream lines around a cylinder through addition of circulation.

from the illustration, the points of stagnation will be displaced downward, while the velocities will be augmented on the upper side and reduced accordingly on the lower. The degree of displacement of the stagnation points will depend upon the ratio of the tangential velocity due to circulation, $v_c = \Gamma/2\pi r_0$, to the velocity of the oncoming flow v_0; the larger this ratio, the greater the displacement.

The form of this complex function will then be:

$$w = v_0 \left(z + \frac{r_0^2}{z} \right) + \frac{i\,\Gamma}{2\pi} \ln z$$

from which,

$$\phi = v_0\, x \left(1 + \frac{r_0^2}{x^2 + y^2} \right) - \frac{\Gamma}{2\pi}\, \theta$$

and

$$\psi = v_0\, y \left(1 - \frac{r_0^2}{x^2 + y^2} \right) + \frac{\Gamma}{2\pi} \ln r$$

The tangential velocity around the circumference of the cylinder, which is of particular importance, may be found simply by combining with v_t, the tangential velocity without circulation, the velocity v_c due to the circulatory motion:

$$v = v_t + v_c = 2v_0 \sin\theta + \frac{\Gamma}{2\pi r_0}$$

The two are obviously added on the upper side and subtracted on the lower. It is evident that at the points of stagnation, where the resultant velocity must be zero, the two velocities will be equal and opposite; the ratio of v_c and v_0, therefore, determines the angle of displacement, $\beta = \theta - 180°$:

$$v = 0 = v_t + v_c = 2v_0 \sin\theta + v_c; \qquad \sin\beta = \frac{1}{2} \frac{v_c}{v_0}$$

Thus β will be 30° when the two velocities are equal, 90° when their ratio is equal to 2; as the ratio becomes greater than 2, the point of stagnation will move away from the cylinder, which will then be entirely surrounded by fluid moving in the same angular direction.

Inasmuch as the velocities are higher on the one side of the cylinder and lower on the other, there must be a resultant pressure acting in the positive y direction. The component of force in this direction due to the pressure intensity over an increment of circular arc when integrated entirely around the circle should give the magnitude of this force per unit length:

$$\frac{F}{L} = \int_0^{2\pi} p_d\, r_0 \sin\theta\, d\theta = \frac{\rho}{2} \int_0^{2\pi} (v_0^2 - v^2)\, r_0 \sin\theta\, d\theta$$

$$= \frac{\rho}{2} \int_0^{2\pi} \left[v_0^2 - \left(2v_0 \sin\theta + \frac{\Gamma}{2\pi r_0} \right)^2 \right] r_0 \sin\theta\, d\theta$$

It must again be noted that the two terms are to be added while θ varies through 180°, and subtracted for the remaining 180° of the cycle.

Integration will result in the very pertinent relationship

$$\frac{F}{L} = \rho\, v_0\, \Gamma \tag{59}$$

which states that the force per unit length of cylinder is a direct product of fluid density, velocity of flow, and circulation. The identical expression applies to circulation around bodies of many other shapes. This so-called "Magnus effect" explains in a general manner the principle of the rotor (applied at various times to the propulsion of ships, airplanes, windmills, and so forth), and the reason for the deviation of a spinning ball from its natural trajectory.

26. Successive Transformations. The foregoing conformal transformations involved relationships between only two complex variables, w and z, the flow picture in the z plane being considered in each case a conformal map of the parallel flow in the original w plane. It must be realized, however, that the parallel flow pattern is also a conformal map of the flow net in the z plane in every case, and may be obtained from the latter through the identical functional relationship; that is, if w is a known function of z [$w = f_w\,(z)$], it is obvious that z must then also be a known function of w [$z = f_z\,(w)$]. Moreover, beginning with the original flow picture, any number of transformations may be performed in succession, each operation resulting in a new conformal map of the preceding flow net, all such maps then being related through the series of successive functional operations.

In performing more than one conformal transformation it is most expedient to introduce an additional plane for the complex variable ζ (zeta) $= \xi + i\,\eta$, in which the ξ (xi) axis is real and the η (eta) axis is imaginary. If now z is any known function of ζ, and w in turn is some known function of z, then w is also a function of ζ, which may be written as follows:

$$w = f_w\,(z); \qquad z = f_z\,(\zeta); \qquad w = f_w\,[f_z\,(\zeta)] = f'_w\,(\zeta)$$

Substituting the real and imaginary parts of the several complex variables, this becomes:

$$\phi + i\,\psi = f_w\,(x + i\,y) = f'_w\,(\xi + i\,\eta)$$

Hence, the real variables ϕ and ψ must be functions of the real variables ξ and η, so that the ζ plane will again represent a transformed flow net of ϕ and ψ lines.

Since further transformation of the picture of flow around a cylinder offers a wide variety of useful flow nets, it is customary to accept the transformation from the w plane to the z plane as performed once and for all, and to proceed at once from the flow picture as shown in Fig. 53a. This procedure is based upon the methods of conformal mapping already discussed at length, and will simply be outlined in the following pages.

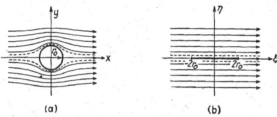

FIG. 53.—Transformation from flow around a cylinder to flow along a thin plate.

If the functional relationship between the z and ζ planes is such that

$$\zeta = v_0\left(z + \frac{r_0^2}{z}\right) \quad \text{or} \quad z = \frac{\zeta \pm \sqrt{\zeta^2 - 4 v_0^2 r_0^2}}{2v_0}$$

it is evident that the conformal map of the flow around the cylinder will become simply a case of parallel flow identical with that in the original w plane. The operation will, however, flatten the circular cross section of the cylinder into an ellipse having a major axis equal to twice the diameter of the circle and a minor (or vertical) axis with a length of zero; that is, the body immersed in the parallel flow is bounded by two straight lines, both of which are superposed upon the real axis of the ζ plane. This flow net is simply that of movement along an infinitely thin plate parallel to the flow (refer to Fig. 53b).

If, on the other hand, the function given above is multiplied by the imaginary factor i,

$$\zeta = i v_0\left(z + \frac{r_0^2}{z}\right)$$

the conformal map of flow around a cylinder will be that of flow around an infinitely thin plate that is normal to the oncoming fluid, as shown in Fig. 54a. If these two operations are combined in the form

$$\zeta = (a + ib) v_0 \left(z + \frac{r_0^2}{z}\right) = a v_0 \left(z + \frac{r_0^2}{z}\right) + i b v_0 \left(z + \frac{r_0^2}{z}\right)$$

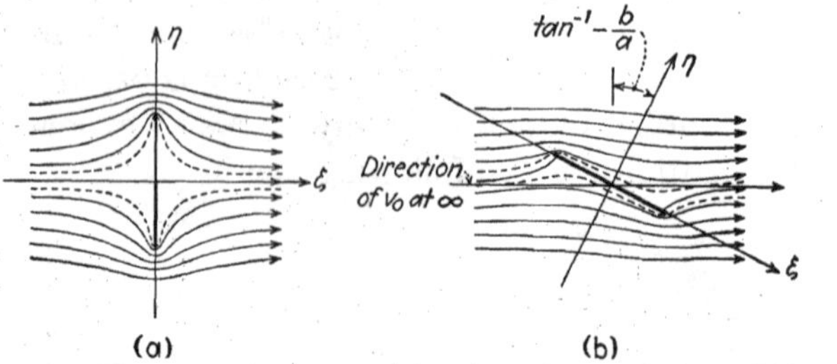

FIG. 54.—Change in flow pattern resulting from rotation of plate.

variation of the constants a and b will change the slope of the plate (see Fig. 54b) and at the same time change the scale of the general picture. This operation will be clear after reference to the elementary principle of rotating the coordinate axes, as illustrated in Fig. 43.

It will be apparent from inspection of Fig. 54b that the unsymmetrical location of the two stagnation points will result in a

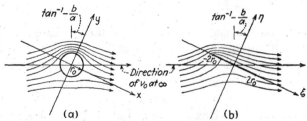

FIG. 55.—Introduction of circulation to eliminate infinite velocity at trailing edge.

force couple tending to rotate the plate about its center of gravity (the coordinate origin), although in such potential motion there can exist no resultant force that would cause a displacement of the center of gravity itself. In order to produce such a force in one direction or another, it is necessary to introduce a constant circulation about the original cylinder, as shown in Fig. 55a. If the above-given functional relationship between ζ and z still exists, the picture of flow in the ζ plane will now represent a combination of parallel flow and circulation around a flat plate inclined at any angle, the force acting upon the plate in a direction normal to the oncoming flow again being given by Eq. 59.

Since the edges of an infinitely thin plate have a zero radius of curvature, it is evident that in general the velocity at such points must be infinitely great. One such region, however, may be obviated in the following manner. Since the position of the two points of stagnation of the circumference of the cylinder may be controlled through varying v_0 and Γ, and since the angle of inclination of the plate to the oncoming flow may also be controlled through variation of the constants a and b, it is possible either to change Γ or to rotate the coordinate axes in Fig. 55a through such an angle that one point of stagnation will coincide with the intersection of the real axis and the circular cross section of the cylinder. Under these conditions in the transformed picture one stagnation point must lie at one end of the thin plate—that is, the stream line at this point will be a smooth continuation of the line of the plate itself, as shown in Fig. 55b.

27. Kutta and Joukowsky Profiles. If the coordinate axes in the picture of flow around a cylinder are transposed bodily in the negative y direction, application of the foregoing transformation will result in the following interesting picture: In Fig. 56 the original section is shown as a heavy circle, the real axis having been displaced downward a distance f. The lighter circle has its center at the new coordinate origin, both circles crossing the real axis at the same two points. It is evident that the operation which transforms the lighter circle into a double line lying upon the real axis in the ζ plane will affect the heavy circle in the same way only at the two points common to both. Every other point on the heavy circle will be shifted upward through the transformation by an amount varying with the distance from the imaginary axis; that is, since points a and b are transformed to $a'b'$ at the real axis, the fact that c and d lie above a and b, respectively, signifies that $c'd'$ must also lie above $a'b'$. The two points on the y axis will finally lie the distance $2f$ above the origin, and the conformal map of the heavy circle will be a double circular arc passing through these three points, as drawn in Fig. 56. Obviously, all stream and potential lines of flow around the heavy circle will undergo a corresponding transformation, so that the final flow net will represent potential motion around a plate bent in the form of a circular arc, called the Kutta profile.

The addition of circulation to the picture will then produce a force upon the curved plate at right angles to the oncoming flow, and by rotating the coordinate axes, in addition to transposing them, the point of stagnation may be moved to the rear edge of the plate as the plate is inclined to the flow.

Had the coordinate axes been shifted to the right, instead of downward, the conformal map of the original circle would have been given the form of a streamlined body symmetrical about

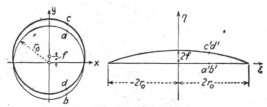

Fig. 56.—The Kutta transformation.

the real axis. That this is true will be seen from Fig. 57; since the heavy circle lies outside of the one concentric with the new coordinate origin at all points but one, then the transformed image of the heavy circle must lie outside of the image of the other (in this case the image is a straight line) at all points except the one common to them both. If the axes are first displaced and then rotated through a given angle, the transformed image will be that of a symmetrical streamlined body of infinite length

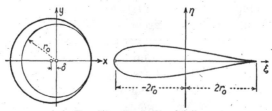

Fig. 57.—Streamlined foil.

(*i.e.*, infinite in a direction normal to the plane of motion) inclined at the given angle to the direction of oncoming flow. Circulation may then be introduced and the rear point of stagnation brought to the trailing edge of the body in the manner already described. It is evident that through this general transformation both regions of infinite velocity have been eliminated.

A combination of these two types of displacement of the coordinate axes—downward and to the right simultaneously—

will result, obviously, in a form of profile that is a combination of the circular arc of the first example with the symmetrical streamlined section of the second example. As shown in Fig. 58, the image of the original section (both image and original are shown in heavy lines) must have in general the same position relative to the other two images as the heavy circle has to the other two original circles—that is, where the heavy circle lies between the other two, its image must also lie between the arc and the straight line; where the heavy circle lies outside the others, its image must also lie outside. Careful study of the illustration will serve to clarify this point.

It will be evident to the reader that this form of section, called the Joukowsky profile after the scientist who developed the trans-

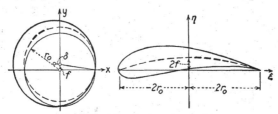

FIG. 58.—The Joukowsky profile.

formation, bears a striking resemblance to the airfoil cross section used for airplane wings and propeller and turbine blades of various sorts. Since the Joukowsky profile may be given almost any thickness and shape through varying f and δ (refer to Fig. 58), and since the velocity and pressure distribution in the surrounding fluid may be found analytically by means of the proper functional relationships, the method has been of great aid in the study of actual airfoil sections. The foil may not only be inclined at any desired angle to the oncoming flow, but through the introduction of circulation the flow net may be made to pass the trailing edge without a singular point of infinitely high velocity; and since the leading edge is rounded, the singular point of the Kutta profile is entirely obviated. Under these conditions the magnitude of the lift, or force normal to the oncoming flow, may be computed from Eq. (59).

Although the problem of velocity distribution around such a profile does not warrant further discussion in this text, a simple graphical method of determining the form of the profile itself may be of interest to the reader. From the origin of the ζ plane the distance $\eta = f$ is laid off along the imaginary axis. Through this point and the point $\xi = r_0$ on the real axis is drawn a straight line, along which, on either side of the imaginary axis, is measured the distance δ, as shown in Fig. 59. Using these two points as centers, two circles are described, both of which pass through the intersection of the straight line with the real axis. Two lines are now drawn from the origin, making equal positive and negative angles with the real axis; from the intersection of one line with the larger circle and from that of the other line with the small circle two more lines are drawn to form a parallelogram. Since the corner of this parallelogram opposite the coordinate origin determines one point on the profile, the operation may be performed simply by swinging two circular arcs having radii equal respectively to the two sides already drawn. This operation is then repeated for as many points as are necessary to complete the profile. Obviously, if the distance δ is equal to zero, the foil will have the form of a circular arc— the Kutta profile—the degree of curvature depending upon the relative magnitude of f and r_0; similarly, if f is zero, there will result a symmetrical streamlined section, its relative thickness depending upon the ratio between δ and r_0.

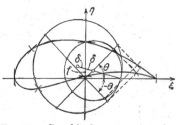

FIG. 59.—Graphical construction of a Joukowsky foil.

28. Methods of Application. In a few specific instances the several elementary transformations already discussed may be applied directly to problems of fluid motion. That such cases are not more numerous is due in part to the restrictions noted in the following chapter, but also to the fact that these elementary transformations are definitely limited in scope. Nevertheless, it is sometimes possible to combine a number of basic functions— namely, those for parallel flow, circulation, and sources and sinks —in such a way as to approximate the desired conditions of motion to a satisfactory degree. This method has been used, for instance, by Spannhake[1] in the study of flow around the impeller

---

[1] SPANNHAKE, W., "Neue Darstellung der Potentialströmung durch Kreiselräder für beliebige Schaufelform," Vorträge aus dem Gebiete der Aerodynamik und verwandter Gebiete (Aachen, 1929), Springer, Berlin, 1930.

blades of a centrifugal pump. In general, however, it must be said that a procedure such as this requires considerable persistence and mathematical ingenuity, and becomes the more difficult as the boundary conditions increase in complexity.

Mention has already been made of the fact that a flow net may be fitted graphically to any two-dimensional boundary profile, however involved the form may be. Justification for this procedure is embodied in the Cauchy integral theorem,[1] which states, in effect, that any pattern of potential motion is determined uniquely by the boundary geometry. The Cauchy integral formula, moreover, provides a means of obtaining this pattern of motion for any two-dimensional boundary conditions whatever, by relating the complex function for any point in the flow to that of a point moving around the boundary profile.[2] Although the calculation must proceed by means of graphical integration, the resulting solution is far more accurate than that obtained by graphical adjustment of the flow net.

This process is very laborious, however, and certain mechanical substitutes for the mathematical analysis will yield the same results with far less trouble. Principal among these is the electrical analogy, which warrants a brief explanation at this point. It has long been known that the potential patterns used herein to describe fluid motion apply fully as well to magnetic and electrical fields. Indeed, the lines of constant velocity potential for a given flow pattern would coincide with lines of constant electrical potential were a current to be passed through a plane conductor having the same boundary outlines. This fact at once suggests the use of the analogy between the two cases as a method of obtaining electrically the flow pattern for any desired boundary conditions. The conducting medium is given the required form, a known potential is applied between the two extremities, and the drop in potential is measured at suitable points over the profile; by selecting equal potential increments, successive lines of constant potential can readily be traced. In some instances a metallic conductor is desirable, although use of an electrolyte is often advantageous. In the latter instance, since the boundaries are formed by flexible strips immersed in the bath and curved to the required shape, the potential may be applied

[1] DURAND, W. F., Mathematical Aids, "Aerodynamic Theory," vol. 1, p. 10, Springer, Berlin, 1934.
[2] *Ibid.*, p. 11.

between either pair of opposite boundaries by proper choice of conducting and non-conducting material. If the outermost stream lines are formed by metallic strips, and the end potential lines by strips of an insulating material, the drop in electrical potential will then indicate the change in ψ from one stream line to the next. That is, the method will permit the location of both stream and potential lines, since in any conformal net the terms ϕ and ψ may be interchanged without affecting the mathematical nature of the pattern. Further information on apparatus and experimental technique may be found elsewhere.[1]

As yet no attempt has been made in this chapter to treat cases of motion which are not determined by fixed boundaries. Since the outermost stream lines invariably coincided with the boundary profiles, no restriction had to be made as to the boundary pressure distribution, the latter following from the existing distribution of velocity. A free surface, on the other hand, not only denotes a discontinuity of the fluid medium, but also represents a stream line along which the pressure intensity must be constant. Needless to say, the Cauchy theorem applies as well to such cases of motion, although if the action of fluid weight can still be neglected, other methods will prove more useful. For instance, it has been found helpful in conformal mapping to introduce an intermediate transformation to the v plane [refer to Eqs. (57) and 58)], in which the pattern—known as a hodograph[2]—indicates by the magnitudes of r and θ the magnitude and direction of the velocity vector at every point of the flow profile in the z plane. By similar means Helmholtz and Kirchhoff first investigated the problem of jet contraction,[3] while von Mises[4] determined the contraction and discharge coefficients for a wide variety of orifice forms.

[1] HOHENEMSER, K., Experimentelle Lösung ebener Potentialströmung, *Forschung auf dem Gebiete des Ingenieurwesens*, vol. 2, no. 10, p. 370, 1931. For an extensive bibliography see HAGUE, B., *The Electrician*, vol, 102, pp. 185, 315, 1929.

[2] PRANDTL-TIETJENS, "Fundamentals of Hydro- and Aerodynamics," p. 178. See also BETZ, A., and PETERSOHN, E., Anwendung der Theorie der freien Strahlen, *Ingenieur-Archiv*, vol. 2, 1931.

[3] LAMB, "Hydrodynamics," p. 94.

[4] MISES, R. VON, Berechnung von Ausfluss- und Ueberfallzahlen, *Zeitschrift VDI*, p. 47, 1917. See also SCHACH, W., Umlenkung eines freien Flüssigkeitsstrahles an einer ebenen Platte, *Ingenieur-Archiv*, vol. 4, 1934; vol. 6, 1935.

Once fluid weight plays an essential role, the hodograph method is not always sufficient, although the Cauchy integral formula may still be counted upon to yield satisfactory results. The approach is now indirect, however, for application of the formula requires prior knowledge of the boundary form, whereas the form of a free surface is the principal variable to be determined. Nevertheless, the formula will yield a flow net for any assumed boundaries, and it is, therefore, only necessary to make a reasonable assumption and then check the resulting solution for constancy of pressure along the assumed free surface. This may be accomplished in the following manner: Since along the free surface

$$v = \sqrt{2g\ (E_w - h)} = \frac{\partial \phi}{\partial s}$$

the magnitude of ϕ at any point on this stream line may be found by integrating v along s from some arbitrary point of reference s_0:

$$\phi = \int_{s_0}^{s} \sqrt{2g\ (E_w - h)}\ ds \tag{60}$$

Comparison of this distribution of ϕ with that obtained by application of the Cauchy formula will show at once the manner in which the assumed profile must be adjusted for the next approximate solution. Evidently, successive trials will lead closer and closer to the exact profile form, three or four approximations generally being sufficient. Unfortunately, space does not permit more detailed discussion of this method, in particular since it has not yet received the attention which it merits. A complete description may be found in a paper by Lauck,[1] who determined therewith the profile of flow over a weir of infinite height.

Since application of the electrical analogy makes recourse to the Cauchy theorem unnecessary in dealing with problems of confined flow, it should also be useful in cases of flow with a free surface—although the possibility of such application apparently

---

[1] Lauck, A., Ueberfall über ein Wehr, Z. angew. Math. Mech., vol. 5, p. 1, 1925.

has escaped the notice of the research world. The process of successive approximation would then proceed in the manner just outlined, the electrical method replacing the Cauchy formula in obtaining the distribution of ϕ to check against that determined by means of Eq. (60). Unfortunately, the hydraulician who is still so optimistic as to seek a simple mathematical method of expressing the form of surface profiles will doubtless continue to seek in vain; no other means of analysis is known for curvilinear flow under gravitational action.

APPLICABILITY OF THE FUNDAMENTAL EQUATIONS

Résumé. Classical hydrodynamics, in the course of its two centuries of development, centered its attention upon the interplay of velocity and dynamic pressure for a given fluid density and given boundary geometry. Although the resulting equations of motion pertain to the case of confined flow, introduction of a single force property—specific weight—readily permits extension of these equations to the case of flow with a free surface. The foregoing chapters have placed considerable emphasis upon the more essential concepts of classical hydrodynamics, since these basic equations are the foundation of modern fluid mechanics. Thus, in Chapter II the elementary principles of momentum, energy, and continuity were discussed in their relationship to the general pattern of motion. Chapter III inquired more thoroughly into the behavior of the fluid particle, stressing the kinematic—rather than the dynamic—aspects of flow. Then, under the assumption of steady, irrotational motion (a restriction that is in itself purely kinematic), certain methods were outlined in Chapters IV and V, permitting the determination of the flow pattern for given boundary conditions—namely, the concept of potential flow, conformal mapping, and the Cauchy integral theorem; the flow net may be looked upon as the graphical representation of the latter principles.

In addition to providing a preliminary structure for the refinements of fluid mechanics, in many instances these basic concepts are directly applicable to engineering problems without further modification. Moreover, the elementary principle of energy provides a qualitative check upon their limit of applicability. This limit depends, obviously, upon the extent to which the original premises of classical hydrodynamics are actually fulfilled in the given problem. Thus, if the principle of potential motion is accepted as the most useful immediate tool provided by the science, it is evident that any flow studied by this means must be essentially one of constant energy, constant density, and complete conformity with the boundary geometry.

Energy Criteria. Whenever flow proceeds from a state of rest, the fluid energy will at first be uniformly distributed; it is then only reasonable to presume that the flow net will satisfactorily describe the pattern of motion. For instance, the discharge under a sluice gate leading from a large reservoir will differ only imperceptibly from potential motion. As will be seen in Part Two of this book, however, the energy of flow will change with distance in the direction of motion, varying not only along each stream line but from one stream line to the next. At low velocities the rate of change of energy in the longitudinal direction will be small, but the energy will then vary appreciably over a normal section. At high velocities, on the other hand, the energy will be fairly uniform across the flow, while the rate of change in the direction of motion will be relatively large. It so happens that the latter type of motion is most often encountered in engineering practice.

In problems of rapid variation in velocity, indeed, the distribution of dynamic pressure is of paramount importance, interest in energy variation then being completely secondary. If the boundary transition is short, the longitudinal energy change will often be of negligible magnitude, and if, in addition, the velocity is sufficiently high, the energy will also be approximately uniform over the cross section of flow. One might therefore conclude that use of the flow net would then be justifiable—at least so far as energy criteria are concerned—regardless of whether the fluid accelerates or decelerates as a result of the boundary form. Nevertheless, the remaining two restrictions—constant density and conformity with the boundary—often limit the extent to which rapid variation may occur if the flow net is to provide satisfactory results.

Variation in Density. An extreme case of rapid transition is shown in the case of a thin flat plate in a plane normal to the direction of motion (Fig. 54a). Although the fluid comes to rest at the point of stagnation, with an accompanying increase in the dynamic pressure, at the edges of the plate the acceleration is infinitely great—a condition requiring a velocity of positive infinity and a pressure intensity of negative infinity. Such conditions are mathematically possible in potential flow, but since they are physically quite out of the question, it is evident that the flow net cannot possibly describe the actual state of motion past such a boundary. '

Since a change in velocity must always be accompanied by a change in dynamic pressure, it is evident that either acceleration or deceleration must tend to change the density of the moving fluid. The reader will realize that in the case of a gas, which is readily compressible, the flow net will have quantitative significance only if the relative change in density is limited to a very small magnitude. Otherwise, the actual flow pattern will differ from that of potential motion in the distribution of both pressure and velocity. Of distinct value, therefore, would be an approximate relationship for the permissible variation in velocity or pressure whereby change in density will not seriously affect such computations.

The differential equation of energy for steady flow applies just as well to a gas as to a liquid, provided the density is no longer treated as a constant; thus,

$$\rho v \, dv + dp + d(\gamma h) = 0$$

Since the increment $d(\gamma h)$ is ordinarily insignificant in gaseous motion, it may be omitted without appreciable error. Dividing by ρ, and integrating along a stream line between points 1 and 2,

$$\frac{v_2{}^2 - v_1{}^2}{2} + \int_{p_1}^{p_2} \frac{dp}{\rho} = 0$$

Although the first term is identical with that for liquid motion, the second differs in that the integration cannot be performed until the two variables are related to one another. This relationship is found in the thermodynamic principle

$$(\text{absolute pressure}) \times (\text{specific volume})^k = \text{constant}$$

Since specific volume is defined as volume per unit weight, it is seen to be the reciprocal of specific weight, whence

$$\frac{p}{\gamma^k} = \text{constant} = \frac{p_1}{\gamma_1{}^k}$$

and therefore

$$\frac{p}{\rho^k} = \frac{p_1}{\rho_1{}^k}$$

or, solving for ρ,

$$\rho = \rho_1 \left(\frac{p}{p_1}\right)^{\frac{1}{k}}$$

The second term of the energy equation may now be rewritten in the form,

$$\int_{p_1}^{p_2} \frac{dp}{\rho} = \frac{p_1^{\frac{1}{k}}}{\rho_1} \int_{p_1}^{p_2} p^{-\frac{1}{k}} dp = \frac{p_1}{\rho_1} \frac{k}{k-1} \left[\left(\frac{p_2}{p_1}\right)^{\frac{k-1}{k}} - 1\right]$$

introduction of which in the energy equation then yields,

$$\frac{v_2^2 - v_1^2}{2} + \frac{p_1}{\rho_1} \frac{k}{k-1} \left[\left(\frac{p_2}{p_1}\right)^{\frac{k-1}{k}} - 1\right] = 0$$

This expression represents the counterpart of the Bernoulli equation for a gas under adiabatic conditions.

Solving for p_2,

$$p_2 = p_1 \left[1 + \frac{\rho_1}{p_1} \frac{v_1^2 - v_2^2}{2} \frac{k-1}{k}\right]^{\frac{k}{k-1}}$$

Subtracting p_1 from both sides of the equation, and expanding the right side in a series,

$$p_2 - p_1 = p_1 \left[1 + \frac{\rho_1}{p_1} \frac{v_1^2 - v_2^2}{2} + \frac{1}{2k}\left(\frac{\rho_1}{p_1} \frac{v_1^2 - v_2^2}{2}\right)^2 + \cdots - 1\right]$$

Finally, writing the pressure and velocity differences in simplified form,

$$\Delta p = \rho_1 \frac{\Delta(v^2)}{2} \left[1 + \frac{\rho_1 \Delta(v^2)}{4kp_1} + \cdots \right]$$

Immediately apparent is the fact that the first term of the series corresponds to conditions of flow at constant density. It then follows that for relatively small changes in velocity the per cent error in pressure change caused by assuming the density constant will be indicated approximately by the magnitude of the quantity

$$\frac{\rho \Delta(v^2)}{4kp} \times 100$$

For instance, in the case of air under normal atmospheric conditions, a change in velocity from 50 to 250 feet per second will result in an error in pressure change of only 1.2 per cent if computed on the basis of constant density. As a matter of fact, the influence of variable density upon the dynamic pattern must be taken into account only under relatively high velocity changes such as are encountered, for instance, in the free efflux of gas from a high-pressure container, or in the motion of airplane propellers and projectiles.

Cavitation. A liquid, unlike a gas, may be considered truly incompressible so far as those types of flow usually encountered in hydraulic engineering are concerned. Nevertheless, while compressive stresses will thus have no further bearing upon the applicability of the flow net in cases of liquid motion, it must be recalled that liquids may not ordinarily be expected to withstand tension to any appreciable degree. In the laboratory, to be sure, it has been shown that water has a tensile strength of at least 34 atmospheres; but it is essential that the water so stressed be extremely clean and free from dissolved air. Under average conditions a liquid will seldom fail to boil once the pressure of vaporization is reached—close to absolute zero at normal temperature.

It is evident that for given boundary conditions either the velocity of flow or the hydrostatic load on a closed system may so be varied that a pressure intensity of absolute zero will be approached at some point on the boundary; that is,

$$p_{abs} = E_v - \rho \frac{v^2}{2} - \gamma h \to 0$$

In the case of the two-dimensional bend of Fig. 11a, the point of minimum pressure intensity will occur along the inner wall near the midpoint of the bend. If now the pressure intensity at this point is reduced to the vapor pressure of the liquid, either by increasing the discharge or by reducing the pressure load, it is evident that the liquid passing this point will just begin to vaporize. If the discharge is further increased, the region of vaporization will grow in size, a cavity forming immediately beyond the midpoint of the bend and appearing to ease the curvature of the innermost filaments. Continued increase in discharge will cause further growth of this vapor pocket, until finally the flow as a whole becomes unstable.

A simple laboratory demonstration of this phenomenon—cavitation—was devised long ago by Osborne Reynolds,[1] and may easily be repeated by the reader. A small glass tube is heated over a flame and drawn out in the form of a Venturi meter

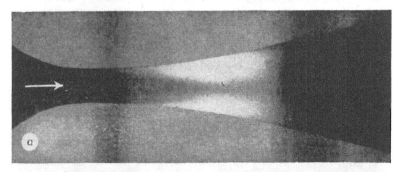

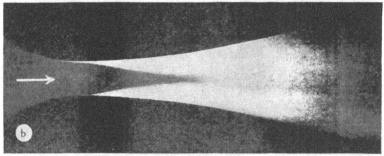

Columbia

FIG. 60.—Cavitation in a two-dimensional Venturi throat; (a) front illumination, (b) rear illumination, under identical conditions of flow.

with a fairly great contraction. If this is connected with a water faucet by means of rubber tubing, at even relatively low rates of flow a white cloud of vapor may be observed just beyond the throat. As the discharge is increased, the zone of cavitation will lengthen, the intensity of the formation and subsequent collapse of the vapor bubbles making itself apparent through a distinct hissing sound and perceptible vibration of the tube. Photographs of a two-dimensional contraction of this nature may be seen in Fig. 60, front illumination emphasizing the region

[1] REYNOLDS, OSBORNE, "Papers on Mechanical and Physical Subjects," vol. 2, p. 578, 1901.

of actual cavitation and illumination from the rear showing the appreciable quantity of air which is brought out of solution by the extreme reduction of pressure and not redissolved. It is quite evident from the photographs that the effect does not extend entirely across the flow section, but is limited to the region of maximum curvature—and hence maximum velocity and minimum pressure—at the boundary.

The occurrence of cavitation in hydraulic machinery is obviously a disadvantage, if only because of the resulting loss in efficiency. There is, however, a far more serious aspect of the problem, the importance of which has led to extensive research on cavitation in this country and abroad. High-speed motion pictures of this process indicate that conditions in the cavitation zone are far from steady—in fact, the vapor cloud seen with the naked eye is merely an average impression, for the successive stages of vaporization, movement downstream, and condensation repeat themselves so many times a second (the number of cycles varying directly with the velocity) that the eye is quite incapable of following. The formation of the vapor pockets is in itself of little consequence—but the abrupt collapse of these cavities as they are carried into regions of higher pressure is accompanied by sudden compressive stresses of exceedingly high magnitude. If the point of collapse is close to a solid boundary, the boundary surface is then subjected to countless intermittent shocks, and will sooner or later fail through fatigue. It was formerly thought that the corrosion—pitting—of metal parts of hydraulic machinery was due to a chemical action intensified by the low pressure. But laboratory tests indicate that such chemically inert (though brittle) substances as glass will fail quite readily, the zone of failure even lying somewhat below the exposed surface. Moreover, in every case the pitting has not been found to occur at the point where the pressure drops, but where the pressure abruptly rises, this marking the region of collapse of the vapor pockets.

Measurements of the pressure distribution along a Venturi throat during cavitation provide an excellent picture of the mean dynamic pattern.[1] In Fig. 61a may be seen a typical series of pressure distribution curves, taken under conditions of constant

---

[1] ACKERET, J., Kavitation, "Handbuch der Experimentalphysik," vol. IV–1, Akademische Verlagsgesellschaft, Leipzig, 1931.

discharge and constant pressure intensity at the entrance, such that the vapor pressure of the liquid would prevail at the point of maximum velocity at the throat. The intensity of cavitation was then governed by varying the downstream pressure, the end of the visible vapor pocket corresponding invariably with the point at which the pressure abruptly began to rise—for example, at point C as shown in the illustration.

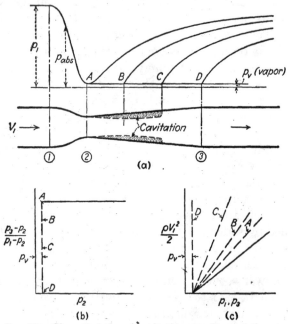

Fig. 61.—Characteristics of cavitation at a Venturi throat.

Since the magnitude of the downstream pressure intensity is seen to determine the intensity of the cavitation phenomenon for the given conditions of flow, one can only conclude that the extent to which the kinetic energy is restored to potential energy beyond the throat will decrease with increasing intensity of cavitation. The plot of relative pressure recovery at section 3 (Fig. 61b) indicates this fact; it will be found, moreover, that if section 3 is moved downstream, points B, C, and D will lie higher on this plot, but will never reach the elevation of point A.[1] It is apparent that the vertical scale then represents the efficiency

[1] Compare MOODY, L. F., and SORENSON, A. E., Progress in Cavitation Research at Princeton University, *Trans. A.S.M.E.*, vol. 57, 1935.

of the meter, cavitation effectively lowering the efficiency of any hydraulic device.

While the foregoing measurements applied to specific pressure and velocity conditions in the approach, they may be generalized in the following manner: Writing the mean energy equation between sections 1 and 2 (ignoring, for convenience, the secondary effects of curvature),

$$\rho \frac{V_1{}^2}{2} + p_1 = \rho \frac{V_2{}^2}{2} + p_2$$

whence

$$p_1 - p_2 = \rho \frac{V_2{}^2 - V_1{}^2}{2} = \rho \frac{V_1{}^2}{2} \left(\frac{V_2{}^2}{V_1{}^2} - 1 \right)$$

The ratio V_2/V_1 is determined by the meter dimensions, and the quantity $\left(\frac{V_2{}^2}{V_1{}^2} - 1 \right)$ may therefore be considered constant for the given meter. Once cavitation begins, p_2 will become equal to p_v, the magnitude of which is also independent of flow conditions. It then follows that p_1 should be a linear function of $\rho V_1{}^2/2$ for all stages of cavitation; that is,

$$p_1 = C \rho \frac{V_1{}^2}{2} + p_v$$

as shown by the full line in Fig. 61c; for any point in the region to the right of the line, cavitation will not occur. Furthermore, for a given magnitude of $\rho V_1{}^2/2$ (and hence of p_1), the degree of cavitation is governed by the pressure intensity at some downstream section. It is then apparent that a given state of cavitation—corresponding, for instance, to the pressure distribution A, B, C, or D—will be indicated by a linear relationship between p_3 and $\rho V_1{}^2/2$:

$$p_3 = C' \rho \frac{V_1{}^2}{2} + p_v$$

The factor C' will vary in magnitude with the stage to which it corresponds, as indicated by the broken lines in Fig. 61c. Moreover, it will remain essentially constant for a given stage of cavitation only if the air content of the water is negligible, for experiments have shown that the characteristics of cavitation will vary with the amount of air in solution.[1]

Cavitation occurs most frequently in two related types of hydraulic machinery—turbine runners and ship propellers—both of which operate under conditions of relatively low pressure. Since as yet no material has been found which resists pitting to a satisfactory degree, the elimination of cavitation remains the only means of solving this costly problem. Three possible courses should be apparent to the reader: reducing the mean velocity of flow, increasing the hydrostatic load, or modifying the curvature at the danger point of the boundary. Since high average velocities and low hydrostatic loads are often essential in the operation of hydraulic machinery at peak efficiency, it is evident that boundary design is of considerable importance. The flow net will be found of considerable assistance in such design, although it is no longer directly applicable once cavitation begins.

Separation. In stating that boundary conditions uniquely determine the form of the corresponding flow net, the Cauchy integral theorem presumes that the outermost stream lines of the resulting flow conform exactly with the boundary profile over their entire length. In other words, neither may cavities exist between the boundary and the fluid medium, nor may a stream line abruptly leave the boundary and wander into the central portion of the flow.

Considering once again the equation of energy along a given stream line,

$$E_v = \rho \frac{v^2}{2} + p + \gamma h = \text{constant}$$

it is evident that for constant values of h a rise in pressure is limited by a zero magnitude of the velocity, just as a rise in velocity is limited by an absolute zero magnitude of the pressure intensity. If the energy is uniformly distributed, the maximum dynamic pressure will then occur at a point of stagnation. But if the energy is different for any two neighboring stream lines, continued increase in pressure in the direction of flow would call for a point of stagnation on one line while the other still displays a finite velocity. Should the stream line of lowest energy lie at a solid boundary, this outermost stream line can no longer

[1] HUNSAKER, J. C., Cavitation Research, *Mech. Eng.*, vol. 57, no. 4, pp. 211–216, 1935.

continue along the boundary once the point of zero velocity is reached, for at a stagnation point the stream line must abruptly change direction.

If the energy equation for any stream line is differentiated (assuming, for convenience, that h is constant), it will be apparent that for a given pressure increment the corresponding change in

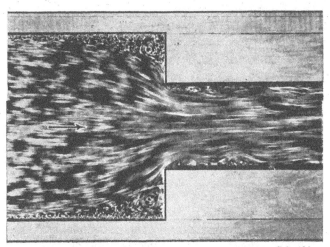

Columbia

FIG. 62.—Abrupt contraction in a conduit, showing separation in regions of local deceleration.

velocity will be inversely proportional to the magnitude of the velocity vector:

$$dv = -\frac{dp}{\rho v}$$

Should the pressure gradient along two neighboring stream lines be essentially the same, the ratio of the corresponding velocity increments will be indicated by the inverse ratio of the velocities:

$$\frac{dv_1}{dv_2} \approx \frac{v_2}{v_1}$$

Evidently, when the pressure decreases (acceleration), the velocity distribution will grow more uniform (Fig. 63a). On the other hand, an increase in pressure (deceleration) will cause the velocities to become more and more unequal, until one eventually reaches the limiting magnitude of zero.

Such circumstances are physically possible only if the point of zero-boundary velocity is a true point of stagnation—that is, the stream line must then abruptly change direction, the flow thereby separating from the boundary, as indicated in Fig. 63b. There thus results a discontinuity in the flow—but not in the fluid, as in the case of cavitation, for the region of discontinuity is generally filled with fluid moving along the boundary in the upstream direction. Since the line of separation nevertheless

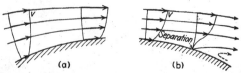

Fig. 63.—Effect of (a) acceleration and (b) deceleration upon velocity distribution.

marks a border of the flow that was not considered in the original boundary conditions, it is obvious that the phenomenon of separation cannot be studied further in the light of potential motion.

While the energy equation permits at least qualitative information as to discontinuity, more thorough discussion of this problem must be left to a later chapter. It must be noted, however, that either local or general deceleration of flow with non-

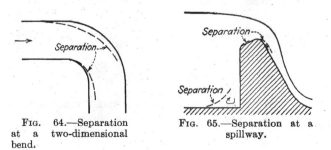

Fig. 64.—Separation at a two-dimensional bend.

Fig. 65.—Separation at a spillway.

uniform energy distribution will almost invariably result in zones of discontinuity. Two such regions are indicated in Fig. 64, conforming with the two-dimensional bend already studied by means of the flow net. Only if the velocity of approach is practically constant from one side to the other, or if the curvature is very gradual, will separation fail to occur. Figure 65, on the other hand, shows a case of generally accelerated motion, the two regions of discontinuity at points of local deceleration near the crest of the spillway being due entirely to poor design. It should be apparent to the reader, therefore, that the flow net attains its greatest significance in the case of rapidly accelerated motion—and, conversely, that a flow profile attains its maximum efficiency when it conforms most closely to conditions of potential motion.

Printed in the USA
CPSIA information can be obtained
at www.ICGtesting.com
LVHW051148131023
760674LV00050B/801